Mathematics

FOR COMMON ENTRANCE

13+

Revision Guide

Mathematics

FOR COMMON ENTRANCE

13+

Revision Guide

Stephen Froggatt
David E Hanson

About the authors

Stephen Froggatt taught Mathematics in Preparatory and Secondary schools for 20 years, during which time he contributed to many websites and articles on teaching Mathematics, and was a regular speaker at Maths teaching conferences. Now he has retired from teaching to pursue a career as a Methodist minister.

David Hanson has over 40 years' experience of teaching and has been Leader of the ISEB 11+ Maths setting team, a member of the 13+ Maths setting team and a member of the ISEB Editorial Endorsement Committee. He has also edited the *SATIPS Maths Broadsheet*. David has retired from teaching to run a small shop trading in collectors' items.

Orders: please contact Bookpoint Ltd, 130 Milton Park, Abingdon, Oxon OX14 4SB. Telephone: (44) 01235 827720. Fax: (44) 01235 400454. Email education@bookpoint.co.uk Lines are open from 9 a.m. to 5 p.m., Monday to Saturday, with a 24-hour message answering service. Visit our website at www.galorepark.co.uk for details of other revision guides for Common Entrance, examination papers and Galore Park publications.

ISBN: 978 1 4718 4689 2

First published in 2015 by

Galore Park Publishing Ltd,

An Hachette UK Company

Carmelite House

50 Victoria Embankment

London EC4Y 0DZ

www.galorepark.co.uk

Impression number 10 9 8 7 6 5 4 3 2

Year 2019 2018 2017

Some illustrations by Ian Moores were re-used. All other illustrations by Aptara, Inc.

Typeset in India

Printed in India

A catalogue record for this title is available from the British Library.

Contents

Introduction

This book has been written to help you revise. It is not a textbook, and the emphasis is on reminding rather than teaching you.

The book contains all the revision material you will need for your Common Entrance exam.

The odd numbered chapters, Number (1), Calculations (1) and so on contain essential reminders of topics covered in KS2.

The even numbered chapters contain reminders of ideas which build upon the earlier foundation.

Chapters 5 and 6 provide you with some practice in using vital skills from all areas of the curriculum which will help you pass your exam.

Throughout you will find plenty of 'Test yourself' questions that will allow you to check that you have learned a section properly, and 'Exam-style questions' with which you practise the kind of questions that you will see in the exam. There are answers near the back of the book.

At the end of each chapter is a summary of what you should have learned. Make sure that you keep track of what you have and have not covered, and keep practising anything you are unsure of.

Examinations are not designed to trip you up, but to give you a chance to demonstrate what you know and what you can do. Make sure that you have the facts at your fingertips so that you can show yourself off at your best.

Good luck!

The syllabus

The ISEB syllabus is revised regularly and is based on the programmes of study for Key Stage 2 and Key Stage 3 of the National Curriculum. The syllabus expects that pupils will be familiar with the skills and knowledge of National Curriculum Key Stage 1 and the National Numeracy Strategy Framework for the early years.

The following table shows how the contents of this book relate to the ISEB syllabus. The purple square indicates material relating to Level 3 only.

Chapter	ISEB syllabus
1 Number (1), 2 Number (2)	
1.1, 2.1 Properties of numbers Multiples (1.1) Factors (1.1) Primes (1.1); prime factorisation (2.1) Squares, cubes and roots (1.1) Odd and even numbers and rules (1.1) Negative numbers (1.1); add, subtract, multiply, divide (2.1) HFC and LCM (2.1)	**Number**
1.2, 2.2 Place value and ordering Reading large numbers (1.2) Multiplying or dividing by 10, 100, 1000 (1.2, 2.2) Ordering numbers (1.2); ordering decimals (2.2)	

1.3, 2.3 Estimation and approximation Rounding numbers (1.3) Significant figures (2.3)	
1.4, 2.4 Fractions, decimals, percentages and ratio Fractions; fractions and decimals (1.4) Decimals and percentages (1.4); fractions as percentages (2.4) Fractions: add, subtract, multiply, divide (2.4) Ratio (1.4); calculating with ratios (2.4)	**Number** **Ratio, proportion and rates of change**
3 Calculations (1), 4 Calculations (2)	
3.1 Number operations	**Number**
3.2 Mental strategies	
3.3, 4.1 Written methods Multiplication and division by a single-digit number (3.3) Multiplication and division to two decimal places (4.1) Long multiplication (alternative methods) (4.1)	
4.2 Calculator methods	
4.3 Checking results Inverses and approximations as checks	
5 Problem solving (1), 6 Problem solving (2)	
5.1, 6.1 Decision making	*Reason mathematically* *Solve problems*
5.2, 6.2 Reasoning about numbers or shapes	
5.3, 6.3 Real-life mathematics	*Develop fluency* *Solve problems*
7 Algebra (1), 8 Algebra (2)	
7.1, 8.1 Equations and formulae Solving quadratic equations (8.1) ■ Solving inequalities and finding integer solution sets (8.1) ■ Algebraic formulae (8.1)	**Algebra**
7.2, 8.2 Sequences and functions Exploring and describing number patterns (7.2) Describing in words the next term of a sequence (8.2) Describing the nth term of a linear sequence (8.2) Quadratic sequences: the nth term and substitution (8.2) ■	
7.3, 8.3 Graphs First quadratic co-ordinates (7.3) Four quadrant co-ordinates (8.3)	
9 Geometry and measures (1), 10 Geometry and measures (2)	
9.1, 10.1 Measures The metric system (9.1) Common units (9.1) Metric and Imperial unit conversions (9.1) Parts of a circle and the value of pi; Pi (10.1) Areas of circles; area and volume formulae (10.1) Calculating areas of fractional and composite shapes (10.1) Calculating volumes of fractional and composite shapes (10.1)	**Geometry and measures**

9.2, 10.2 Shape Drawing solids and nets (9.2) Symmetry and congruence (9.2) Reflection/rotation symmetry in plane shapes (10.2) Types of triangle, types of quadrilateral (9.2) Properties of quadrilaterals (10.2) Polygons (10.2) Drawings of 3D objects on isometric paper (10.2)	
9.3, 10.3 Space Triangle construction; ruler and compass constructions (9.3, 10.3) Types of angle (9.3) The eight-point compass (9.3) Three transformations: reflection, rotation and translation (9.3) Angle laws for straight lines; for polygons (10.3) Area factor for enlargements (10.3) Three-figure bearings and scale drawings (10.3) Pythagoras' theorem (10.3) ■	The sub-topic of bearings and scale drawings also matches to **Ratio, proportion and rates of change**
11 Statistics and probability (1), 12 Statistics and probability (2)	
11.1, 12.1 Statistics Sorting shapes; Venn and Carroll diagrams (11.1) Frequency tables using discrete data (11.1) Finding the mode, median and range (11.1) Calculating the mean; comparing two distributions (11.1) Grouped data and equal class intervals (11.1) Line graphs and conversion graphs (11.1) Drawing conclusions from graphs, charts and diagrams (11.2, 12.2) Frequency diagrams (bar charts) using discrete data (11.2) Pie charts (11.2, 12.2) Scattergraphs and correlation (11.2, 12.2)	**Statistics**
11.2, 12.2 Probability Words to describe probability (11.2) Sets and Venn diagrams (11.2, 12.2) Theoretical and experimental probability (12.2)	**Probability**

Your exam

When sitting the ISEB Common Entrance 13+ exam, you will take three papers:

- a calculator paper lasting one hour (Level 1, 2 or 3)
- a non-calculator paper lasting one hour (Level 1, 2 or 3)
- a mental arithmetic test lasting 30 minutes, which is the same for all levels.

The syllabus requirements for Level 1 and Level 2 are the same, but there are additional requirements for Level 3 (indicated by a tag in this book).

At Level 1, the questions are generally more 'straightforward' and any formulae that are required will be provided.

How to use this book

This book contains symbols that will help you to revise only what you need to.

First, make sure you know which Level of the exam you are taking. If in doubt, ask your teacher.

In the book you will find tags to indicate information or questions relating to the Level you are taking. Look out for these:

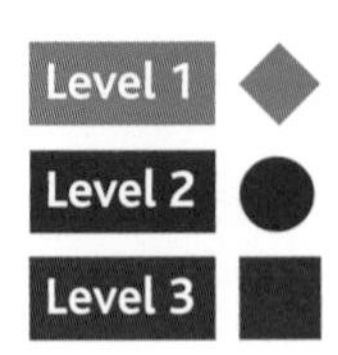

- If you are taking Level 1 or Level 2, you will need to revise everything in the book except for material tagged as Level 3 or 'not examined' (there will be a note in the margin if this is the case).
- If you are taking Level 3, you will need to revise everything in this book except for the few things which will not be examined in CE (such things are indicated by a box in the margin).

The book includes these features to help you with your revision:

Revision tip

The Revision tip boxes contain useful hints on how to revise, and help with remembering key information.

Exam-style questions

At the end of each section, you will find questions in the style of real exam questions to help you practise.

You can find the answers near the end of the book. Exam-style questions have marks allocated, for example (2), as they would in the exam.

★ Make sure you know

Located towards the end of each chapter, these sections contain a summary of the things you really need to know.

Test yourself

At the end of each chapter, you will see a box like this containing questions so you can check that your revision has been successful. Make sure you can answer these questions correctly before moving on to the next chapter. The answers are near the end of the book.

Tips on revising

Get the best out of your brain:

- Give your brain plenty of oxygen by exercising. You can revise effectively if you feel fit and well.
- Eat healthy food while you are revising. Your brain works better when you give it good fuel.

- Think positively. Give your brain positive messages so that it will want to study.
- Keep calm. If your brain is stressed it will not operate effectively.
- Take regular breaks during your study time.
- Get enough sleep. Your brain will carry on sorting out what you have revised while you sleep.

Get the most from your revision

- Do not work for hours without a break. Revise for 20–30 minutes and then take a five-minute break.
- Do good things in your breaks: listen to your favourite music, eat healthy food, drink some water, do some exercise or juggle. Do not read a book, watch TV or play on the computer; it will conflict with what your brain is trying to learn.
- When you go back to your revision, review what you have just learnt.
- Regularly review the facts you have learnt.

Get motivated

- Set yourself some goals and promise yourself a treat when the exams are over.
- Make the most of all the expertise and talent available to you at school and at home. If you do not understand something, ask your teacher to explain.
- Get organised. Find a quiet place to revise and make sure you have all the equipment you need.
- Use yearly and weekly planners to help you organise your time so that you revise all subjects equally. (Available for download from www.galorepark.co.uk)

Know what to expect in the exam

- Use past papers to familiarise yourself with the format of the exam.
- Make sure you understand the language examiners use.

Before the exam

- Have all your equipment and pens ready the night before.
- Make sure you are at your best by getting a good night's sleep before the exam.
- Have a good breakfast in the morning.
- Take some water into the exam if you are allowed.
- Think positively and keep calm.

During the exam

- Have a watch on your desk. Work out how much time you need to allocate to each question and try to stick to it.
- Make sure you read and understand the instructions and rules on the front of the exam paper.

- Allow some time at the start to read and consider the questions carefully before writing anything.
- Read all the questions at least twice. Do not rush into answering before you have a chance to think about it.
- If you find a question is particularly hard, move on to the next one. Go back to it if you have time at the end.
- Check your answers make sense if you have time at the end.

Tips for the Maths exam

- Read the questions extra carefully. It is very easy to miss the word 'not' in a probability question, or the words 'as a percentage' in a number question.
- Show all your working. In a non-calculator question, failure to do this could result in loss of marks; in a calculator question it helps the examiner to check your method even if the final answer is wrong. Candidates often forget that marks can be awarded for working – it is not just about getting the right answer.
- Check that your answer sounds reasonable. A building 7 cm tall, or an average pupil mass of 987 kg, should make you think again. The most common way to get a crazy answer is to forget the units (centimetres and metres in the same question, for example).
- When drawing graphs and doing constructions, use a sharp pencil, and try to be as neat and precise as you can.

1 Number (1)

1.1 Properties of numbers

Types of number

For this section it really helps if you know your tables well! Let us see why they are so important.

The multiplication table

×	1	2	3	4	5	6	7	8	9	10	11	12
1	1	2	3	4	5	6	7	8	9	10	11	12
2	2	4	6	8	10	12	14	16	18	20	22	24
3	3	6	9	12	15	18	21	24	27	30	33	36
4	4	8	12	16	20	24	28	32	36	40	44	48
5	5	10	15	20	25	30	35	40	45	50	55	60
6	6	12	18	24	30	36	42	48	54	60	66	72
7	7	14	21	28	35	42	49	56	63	70	77	84
8	8	16	24	32	40	48	56	64	72	80	88	96
9	9	18	27	36	45	54	63	72	81	90	99	108
10	10	20	30	40	50	60	70	80	90	100	110	120
11	11	22	33	44	55	66	77	88	99	110	121	132
12	12	24	36	48	60	72	84	96	108	120	132	144

You need to know your tables in several ways. First of all, learn them in columns so that you can list, for example, all multiples of seven. Then, be able to jump in at any point, and know that, e.g. $4 \times 8 = 32$, without thinking. Finally, you should be able to use the table backwards and be able to give all factor pairs of a number (e.g. which numbers multiply to give 36?).

Multiples

Multiples of 3 include 3, 6, 9, 12, 15, 18, ...

Multiples of 6 include 6, 12, 18, 24, 30, 36, ...

Think of multiples as the results in the times table of a number.

Factors

Factors of 30 are all the numbers that divide exactly into 30

We can make a factor rainbow.

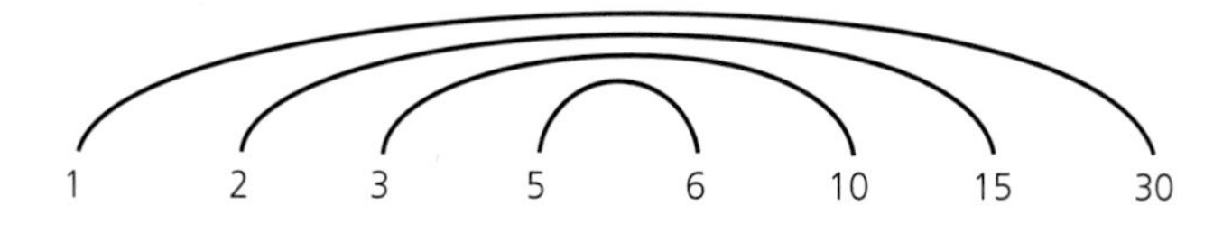

Notice how they come in pairs.

$30 = 1 \times 30$ or 2×15 or 3×10 or 5×6

Think of factors as the *questions* from the times table.

Primes

Quite simply, a prime is a number which has only two factors: 1 and itself.

The first 100 prime numbers

2	3	5	7	11	13	17	19	23	29
31	37	41	43	47	53	59	61	67	71
73	79	83	89	97	101	103	107	109	113
127	131	137	139	149	151	157	163	167	173
179	181	191	193	197	199	211	223	227	229
233	239	241	251	257	263	269	271	277	281
283	293	307	311	313	317	331	337	347	349
353	359	367	373	379	383	389	397	401	409
419	421	431	433	439	443	449	457	461	463
467	479	487	491	499	503	509	521	523	541

Obviously you do not need to learn all these! Try to be familiar with as many as you can, though. At the very least learn the first ten or so, and reach the point where you can recognise all numbers under 100 as either prime or not.

Note: 1 is not prime.

Squares, cubes and roots

Any number multiplied by itself gives a square. In the factor rainbow this number joins to itself in the middle. Let us try this for the number 36:

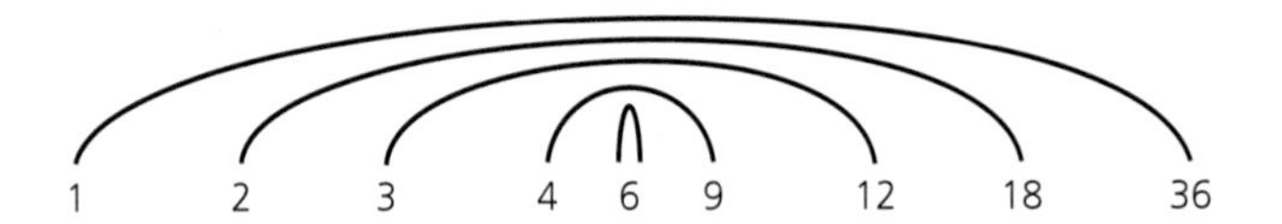

Notice how the 6 is joined to itself because 36 = 6 × 6

Multiply any number (n) by itself and the result is a square number (n^2). Any number multiplied by its square gives a cube number (n^3). It is an extremely good idea to know the squares of at least the first ten numbers and at least the first five cubes. Here are the first 15 of each:

n	n^2	n^3
1	1	1
2	4	8
3	9	27
4	16	64
5	25	125
6	36	216
7	49	343
8	64	512
9	81	729
10	100	1000
11	121	1331
12	144	1728
13	169	2197
14	196	2744
15	225	3375

n^2 means n to the power of 2

2 is the index number.

Similarly, for n^3, 3 is the index number.

Remember the connection between squares and square roots.

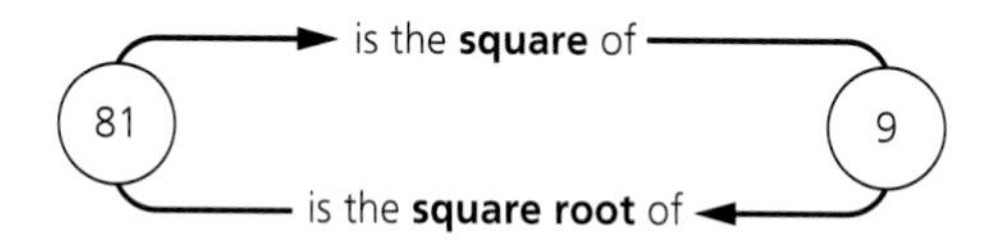

Cubes and cube roots work in a similar way.

Odd and even numbers and rules

There are several general rules about combining two numbers, depending on whether they are odd or even. You may wish to think about why these rules always work as they do.

odd + odd = even
odd + even = odd
even + odd = odd
even + even = even

odd – odd = even
odd – even = odd
even – odd = odd
even – even = even

odd × odd = odd
odd × even = even
even × odd = even
even × even = even

There are no simple rules for division. Any division could be a fraction, which is neither odd nor even.

Negative numbers

Numbers below zero are called negative numbers.

Real-life negatives

Where do we find negative numbers in everyday life? Temperatures are perhaps the most common, but we also talk about negative or overdrawn bank balances (in the red), as well as distances below sea level.

Brr! Below zero

Negative temperatures mean ice and snow. $^{-}3$°C is 3 degrees below zero and so on. Let us extend the number line downwards to see what is going on.

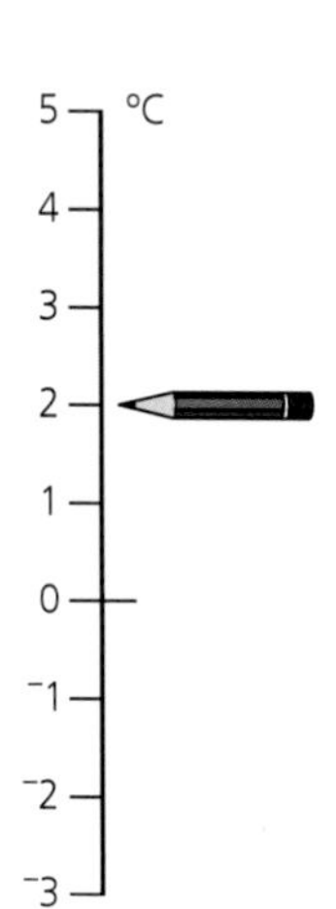

Put your pencil on 2 °C (as shown).

Now move down 3 degrees and you should end up with your pencil on $^{-}1$ °C.

We write: $2 - 3 = {}^{-}1$

Check also that: $5 - 8 = {}^{-}3$

Adding takes us back up again.

${}^{-}2 + 6 = 4$

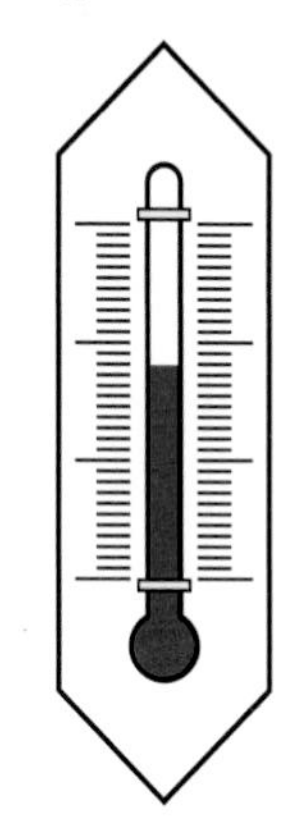

> Adding $^{-}1$ is equivalent to subtracting 1
>
> $5 + {}^{-}8$ is equivalent to $5 - 8 = {}^{-}3$
>
> Subtracting $^{-}1$ is equivalent to adding 1
>
> $10 - {}^{-}4$ is equivalent to $10 + 4 = 14$

Putting negative numbers in order
Remember to think of the number line all the time. That way you will correctly say:

$^{-}15$ is less than $^{-}8$, which is less than 3, which is less than 11, and so on.

Roman numerals

The **Roman numeral** system is an old counting system, which some people believe was based on fingers and hand shapes.

I represents 1 finger

II represents 2 fingers, side by side

III represents 3 fingers side by side

V represents a hand (think of the V shape in the crook of your thumb) so it represents 5 fingers

X represents a V on top of another V, or two hands, or 10 fingers

Other letters are used for large numbers:

L represents 50

C represents 100

D represents 500

M represents 1000

Revision tip

To remember that C = 100 and M = 1000, remember that there are 100 years in a Century and 1000 years in a Millennium.

Revision tip

Try to think up a mnemonic that will help you keep these Roman numerals in order, such as I Value EXcellence: Latin Counters Deserve Medals!

How to read Roman numerals
The simple answer is to add up the letter values.

Examples

XXIII = 10 + 10 + 1 + 1 + 1 = 23

MDCCCLXVII = 1000 + 500 + 100 + 100 + 100 + 50 + 10 + 5 + 1 + 1 = 1867

However, the rules change to **subtraction** when a small value letter appears *before* a large value letter. There are six possible subtraction pairs to remember.

IV = 1 subtracted from 5, giving 5 – 1 = 4

IX = 1 subtracted from 10, or 10 – 1 = 9

XL = 10 subtracted from 50, or 50 – 10 = 40

XC = 10 subtracted from 100, or 100 – 10 = 90

CD = 100 subtracted from 500, or 500 – 100 = 400

CM = 100 subtracted from 1000, or 1000 – 100 = 900

Revision tip

Think to yourself: 1 before 5 is 4, 10 before 50 is 40, and so on.

When we put these two rules together, we first look for any subtraction pairs, and then add up the rest.

MCMLXXX**IV** contains two subtraction pairs: CM (900) and IV (4)

M + CM + L + XXX + IV = 1000 + 900 + 50 + 30 + 4 = 1984

Similarly

MCMXCIX = M + CM + XC + IX = 1000 + 900 + 90 + 9 = 1999

Exam-style questions

Try these questions for yourself. The answers are given near the back of the book.

Some of the questions involve ideas met in earlier work that may not be covered by the notes in this chapter.

1.1 **(a)** Which of these numbers are divisible by 3? (2)

15 26 36 45 56 114 1011

(b) Which of these numbers are multiples of 4? (2)

6 12 16 18 24 36 52

(c) Which of these numbers are factors of 24? (2)

2 3 4 6 8 12 24

1.2 Which of these numbers are prime? (2)

1 2 3 5 9 13 17

1.3 Here are some number cards.

3	4	6	7	12	18

From the cards, choose:

(a) a multiple of 6 (1)

(b) a factor of 6 (1)

(c) a square number (1)

(d) all the prime numbers (2)

(e) the number which is a common multiple of 4 and 6 (1)

(f) two numbers which have a product of 21 (1)

(g) three numbers which have a sum of 25, using each number only once. (2)

1.4 Write the next two numbers in these sequences.

(a) 15 20 25 30 35 ... (1)

(b) 57 54 51 48 45 ... (1)

(c) 17 13 9 5 1 ... (2)

1.5 Write the next two numbers in these sequences.

(a) 1 3 6 10 15 ... (2)

(b) 2 2 4 6 10 ... (2)

(c) 224 112 56 28 14 ... (2)

1.6 Gina is snorkelling.

(a) She dives to $^{-}2$ m (2 metres below the water surface) and looks at a shark 3 metres below her. At what depth is the shark swimming? (1)

(b) On her next dive, Gina swims down to $^{-}2.3$ metres and sees a jellyfish 1.4 metres above her. At what depth is the jellyfish swimming? (2)

1.7 **(a)** At 07:00 one morning, the temperature inside the window was 4 °C and the temperature outside the window was 11 degrees lower. What was the outside temperature? (1)

(b) By 08:30 the temperature outside had risen to $^{-}3$ °C.

By how many degrees had the outside temperature risen? (1)

1.8 Write these numbers written in Roman numerals as ordinary numbers:

(a) CXII (1)

(b) MDCCCLXX (1)

(c) MMXIV (1)

1.9 Write these numbers in Roman numerals:

(a) 57 (1)

(b) 423 (1)

(c) 1789 (1)

1.10 **(a)** Which number, in Roman numerals, is shown on the abacus below? (1)

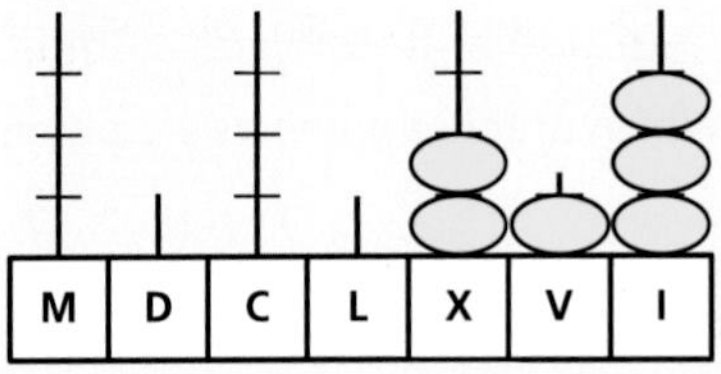

M	D	C	L	X	V	I

(b) Write this number in words. (1)

(c) Write the number in ordinary figures. (1)

(d) On the Roman abacus, why do the D, L and V spikes have space for only one bead? (1)

(e) On this blank Roman abacus draw beads to show the number that is ten times the number shown on the abacus in part (a). (1)

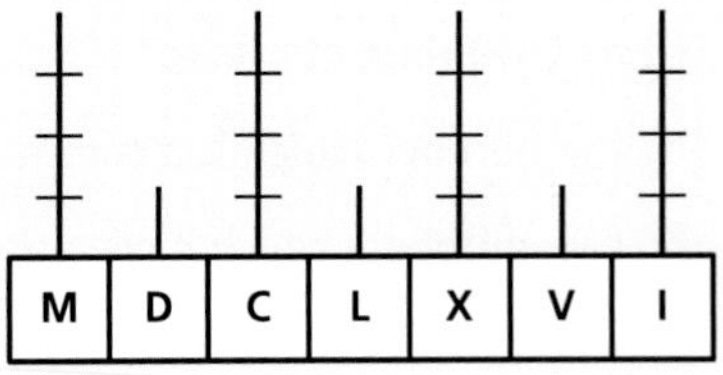

M	D	C	L	X	V	I

1.2 Place value and ordering

Reading large numbers

How do you read 4037021?

How do you write seven million five hundred and nine?

The key is to use commas. Start at the right hand end of the number and mark off the digits in groups of three, using commas:

4,037,021

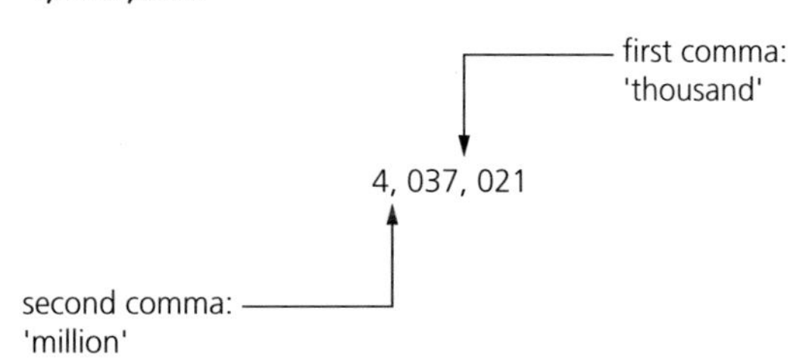

(Additional commas would be needed for even larger numbers. After 'million' the next few commas are called 'billion', 'trillion', 'quadrillion' and 'quintillion' although you don't need to know them for the exam!)

Now read the number aloud in groups of three digits, inserting the appropriate word whenever you reach a comma.

'Four *million*, thirty-seven *thousand* and twenty-one.'

Let us try another. First write 2539684 as 2,539,684. Now read it aloud:

'Two *million*, five hundred and thirty-nine *thousand*, six hundred and eighty-four.'

Working in reverse is similar – replace the words 'million' and 'thousand' by commas. The only thing to watch is the number of zeros:

One *million*, two hundred and eight *thousand* and forty-seven is 1,208,047

Seven *million*, five hundred and nine (note that there are no *thousands*)

is 7,000,509 where we put '000,' to show the 'no thousands'.

Revision tip

Remember that in most books, commas are replaced by spaces. You will often see, for example, 2,539,684 printed as 2 539 684. Write the commas in if you wish!

Multiplying or dividing by 10 or 100

Our number system makes it easy to multiply or divide by ten.

Multiplying by 10 or 100

Multiplying by 10 or 100 is easy. It can be thought of as:

- writing zeros at the end of the *whole* numbers
- moving digits to the left so they have a higher place value
- moving the decimal point to the right.

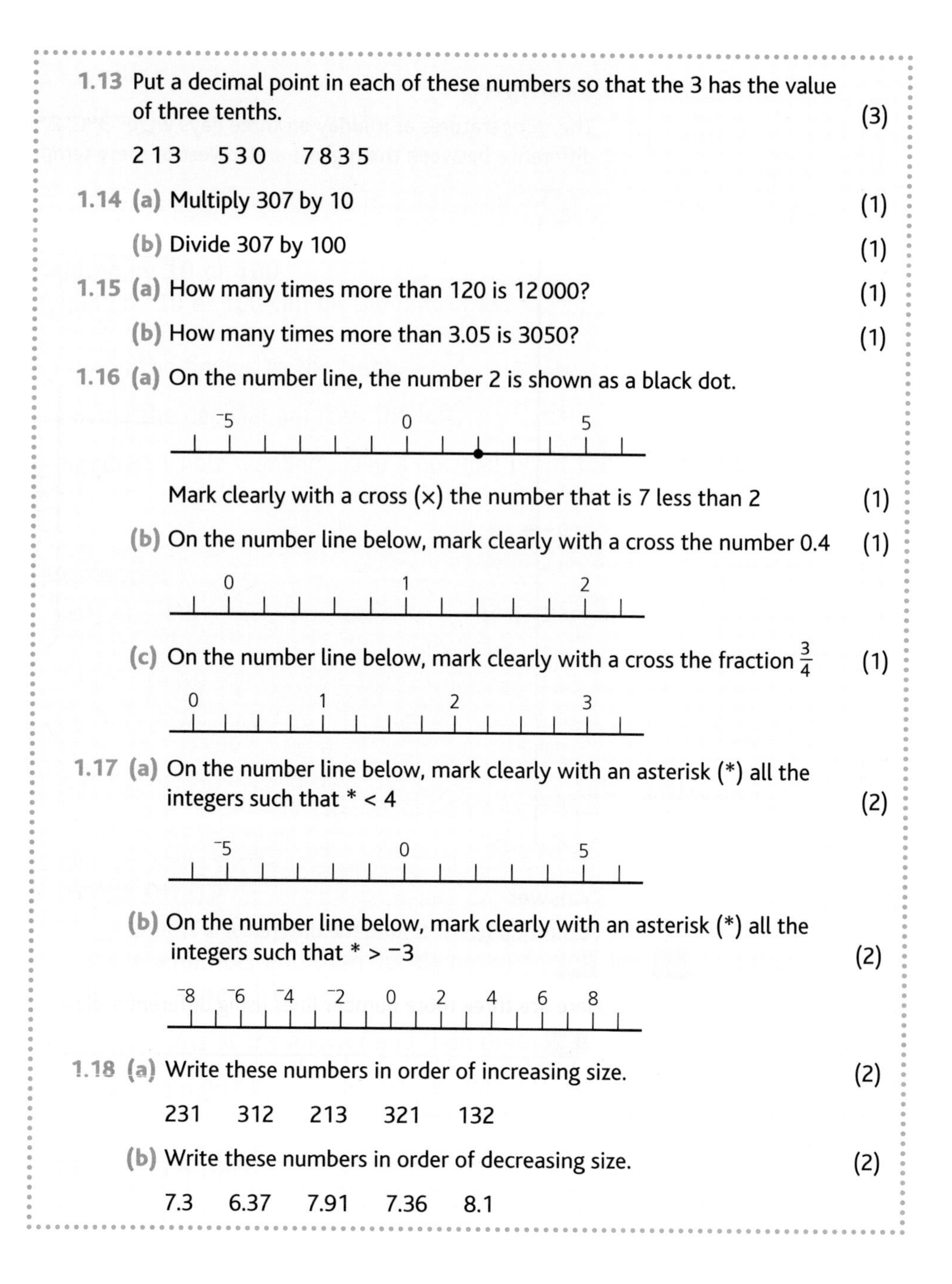

1.13 Put a decimal point in each of these numbers so that the 3 has the value of three tenths. (3)

2 1 3   5 3 0   7 8 3 5

1.14 **(a)** Multiply 307 by 10 (1)

(b) Divide 307 by 100 (1)

1.15 **(a)** How many times more than 120 is 12 000? (1)

(b) How many times more than 3.05 is 3050? (1)

1.16 **(a)** On the number line, the number 2 is shown as a black dot.

Mark clearly with a cross (x) the number that is 7 less than 2 (1)

(b) On the number line below, mark clearly with a cross the number 0.4 (1)

(c) On the number line below, mark clearly with a cross the fraction $\frac{3}{4}$ (1)

1.17 **(a)** On the number line below, mark clearly with an asterisk (*) all the integers such that * < 4 (2)

(b) On the number line below, mark clearly with an asterisk (*) all the integers such that * > −3 (2)

1.18 **(a)** Write these numbers in order of increasing size. (2)

231   312   213   321   132

(b) Write these numbers in order of decreasing size. (2)

7.3   6.37   7.91   7.36   8.1

1.3 Estimation and approximation

Rounding numbers

It is very useful to be able to '**round**' a whole number to give an approximate value.

To help you round, draw a vertical line after the digit you are rounding to. If the number after the line is 5 or more, then round *up*. If the number after the line is 4 or less, then round *down*.

- Round *up*: add 1 to the digit before the vertical line and replace all digits to the right of the vertical line with zeros.
- Round *down*: simply replace all digits to the right of the vertical line with zeros.

Examples

(i)	Round 1327 to the nearest ten	1 3 2|7
	Round up	1 3 3 0
(ii)	Round 3029 to the nearest thousand	3|0 2 9
	Round down	3 0 0 0
(iii)	Round 1250 to the nearest hundred	1 2|5 0
	Round up	1 3 0 0

An abacus can help you to understand rounding.

Example

Round 6435 to the nearest ten.

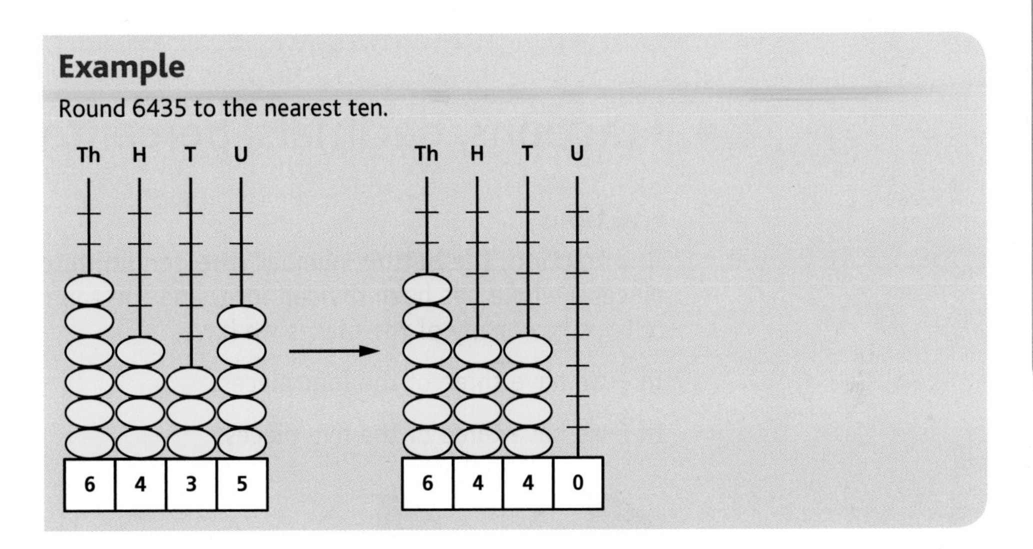

We use the same rules when rounding decimals to the nearest **integer** (whole number). The only difference is that we do not replace the figures to the right of the point with zeros.

Examples

(i)	Round 3.7 to the nearest integer	4
(ii)	Round 4.4 to the nearest integer	4
(iii)	Round 4.5 to the nearest integer	5
(iv)	Round 4.49 to the nearest integer	4

Exam-style questions

Try these questions for yourself. The answers are given near the back of the book.

Some of the questions involve ideas met in earlier work that may not be covered by the notes in this chapter.

1.19 (a) Estimate the position of 3 on this number line. Mark it with a cross. (2)

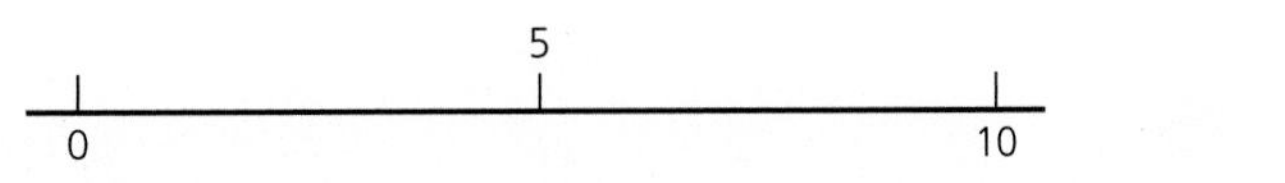

(b) Estimate the position of 60 on this number line. Mark it with a cross. (2)

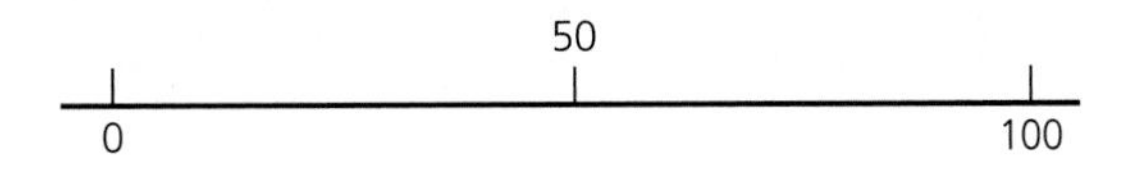

1.20 **(a)** Write 274 to the nearest ten. (1)

(b) Write 751 to the nearest hundred. (1)

(c) Write 9.85 to the nearest whole number. (1)

1.21 **(a)** Estimate (do not calculate exactly) the result of this addition: (1)

2998 + 4003

(b) Estimate the result of this subtraction: (1)

11 002 – 4997

1.22 It has taken Katie about half an hour to read 20 pages of her adventure story book. The book has 405 pages. About how many hours will it take Katie to read the whole book? (3)

1.4 Fractions, decimals, percentages and ratio

Fractions

In a **fraction**, the bottom number (the **denominator**) tells us how many pieces a whole has been divided into, and the top number (the **numerator**) tells us how many of the pieces we have.

In $\frac{3}{4}$ we have three of the four pieces.

In $\frac{3}{5}$ we have three of the five pieces.

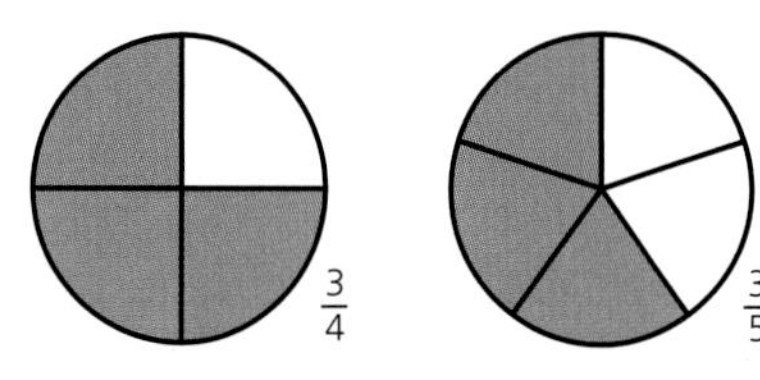

Imagine a cake being cut up. You would eat more cake if you ate three of four pieces than if you ate three of five pieces!

Fractions and decimals

To change a fraction to a decimal, we change it to an **equivalent** fraction with a denominator (bottom number) of 10 or 100

Examples

(i) $\frac{1}{2}$ is the same as $\frac{5}{10}$ 0.5

$\frac{1}{2}$ → (× 5 / × 5) → $\frac{5}{10}$ 0.5

(ii) $\frac{2}{5}$ is the same as $\frac{4}{10}$ 0.4

$\frac{2}{5}$ → (× 2 / × 2) → $\frac{4}{10}$ 0.4

(iii) $\frac{1}{4}$ is the same as $\frac{25}{100}$ 0.25

$\frac{1}{4}$ → (× 25 / × 25) → $\frac{25}{100}$ 0.25

Decimals and percentages

To change a decimal to a percentage, we simply multiply by 100

Examples

(i) 0.5 = 50%

(ii) 0.4 = 40%

(iii) 0.25 = 25%

The fraction wall

A good way to see how fractions and percentages compare is by using the '**fraction wall**':

whole	1 (100%)									
halves	$\frac{1}{2}$	50%								
thirds	$\frac{1}{3}$	$\frac{1}{3}$	33%							
quarters	$\frac{1}{4}$	$\frac{1}{4}$	25%	25%						
fifths	$\frac{1}{5}$	$\frac{1}{5}$	20%	20%	20%					
sixths	$\frac{1}{6}$	$\frac{1}{6}$	$\frac{1}{6}$	$16\frac{2}{3}$%	$16\frac{2}{3}$%	$16\frac{2}{3}$%				
eighths	$\frac{1}{8}$	$\frac{1}{8}$	$\frac{1}{8}$	$\frac{1}{8}$	12.5%	12.5%	12.5%	12.5%		
tenths	$\frac{1}{10}$	$\frac{1}{10}$	$\frac{1}{10}$	$\frac{1}{10}$	$\frac{1}{10}$	10%	10%	10%	10%	10%

Using a ruler held vertically, it is now easy to see that (for example):

$\frac{2}{6} = \frac{1}{3}$ or $\frac{8}{10} > \frac{3}{4}$

Ratio

Fractions (such as $\frac{3}{5}$) compare the number of parts we have (in this case 3) to the total number of parts in the whole (in this case 5).

Ratios can do the same thing, but they can be much more useful.

Look at this group of shapes.

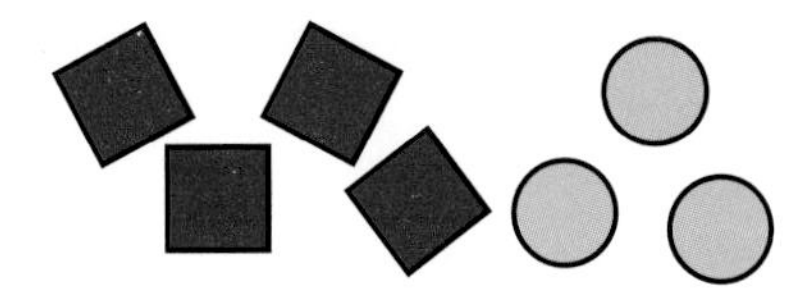

There are four squares and three circles.

The fraction of the shapes that are square is $\frac{4}{7}$

The ratio of the number of squares to the total number of shapes is 4:7 but with ratio, we can go further.

We can also say that the ratio of the number of squares to the number of circles is 4:3

Similarly, the ratio of the number of circles to the number of squares is 3:4

Some fractions can be simplified.

Example

$\frac{6}{9}$ can be simplified to $\frac{2}{3}$. We simply divided top and bottom (i.e. numerator and denominator) by the same number. In this case we divided both by 3

Some ratios can be simplified in the same way.

Example

10:8 can be simplified to 5:4 (dividing both numbers by 2).

Exam-style questions

Try these questions for yourself. The answers are given near the back of the book.

Some of the questions involve ideas met in earlier work that may not be covered by the notes in this chapter.

1.23 In this strip of squares, two of the five have been shaded to show the fraction $\frac{2}{5}$

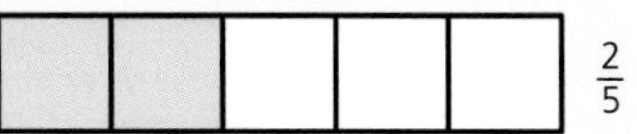

$\frac{2}{5}$

Sketch a similar diagram to show the fraction $\frac{5}{8}$ (2)

1.24 Find $\frac{3}{4}$ of 56 cm. (2)

1.25 Slimy the slug was 60 mm long at the beginning of July. At the end of the month he was $1\frac{1}{4}$ times as long. How long was Slimy at the end of July? (2)

1.26 Gill's aunt gave her £45 and asked her to save two-thirds of it. How much did Gill save? (2)

1.27 (a) Which is bigger, $\frac{1}{6}$ or $\frac{1}{7}$? (1)

(b) Which is bigger, $\frac{3}{4}$ or $\frac{2}{3}$? (1)

1.28 Lachlan has drawn a machine that finds equivalent fractions.

$\frac{2}{3}$ → × 4 / × 4 → $\frac{8}{12}$

(a) Fill in the details to show how the machine can change $\frac{1}{2}$ into sixths. (2)

$\frac{1}{2}$

(b) Fill in the details to show how the machine can change $\frac{3}{4}$ to twelfths. (2)

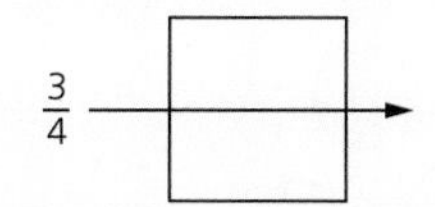

$\frac{3}{4}$

1.29 Complete the table, showing the equivalent fractions, decimals and percentages. (9)

Fraction (in simplest form)	$\frac{1}{2}$				$\frac{3}{10}$
Decimal		0.75		0.7	
Percentage	50%		40%		

1.30 Sally has five cats and two dogs.

(a) What is the ratio of the number of cats to the number of dogs? Give your answer in the form _ : _ . (2)

(b) What is the ratio of the number of dogs to the total number of pets? (2)

★ Make sure you know

- ★ All of the material in National Curriculum Key Stage 1; for example, rounding integers to the nearest 10 or 100
- ★ How to recognise and describe number patterns and relationships, including multiples, factors and squares
- ★ What is meant by a prime number, and why 2 is prime, but 1 is not
- ★ How to order, add and subtract negative numbers in context
- ★ How to use your understanding of place value to multiply and divide whole numbers by 10 or 100
- ★ How to round numbers efficiently
- ★ How to recognise approximate proportions of a whole and be able to use simple fractions and percentages to describe these
- ★ How to read Roman numerals to 1000 (M) and recognise years written in Roman numerals
- ★ How to use the glossary at the back of the book for definitions of key words

Test yourself ✓

Before moving on to the next chapter, make sure you can answer these questions.

The answers are near the back of the book.

1 Here is a number grid.

1	2	3	4	5	6
7	8	9	10	11	12
13	14	15	16	17	18
19	20	21	22	23	24
25	26	27	28	29	30
31	32	33	34	35	36

(a) Circle 3 and then every third number.

(b) Put a cross on 4 and then every fourth number.

(c) What can you say about the numbers that are circled and also have a cross?

2 Here is part of a multiplication square that has been partially completed.

×	2	3	4	5	6	7	8
2				10			
3	6						
4							
5		15					
6							
7							
8							

(a) Carefully shade in all the multiples of 9

(b) Put a circle around all the square numbers.

(c) If the square were completed, which numbers would appear most often?

3 **(a)** What is a prime number?

(b) Without looking at the table in Section 1.1 write, in order, the prime numbers up to at least 30

4 Jasmine has three number cards.

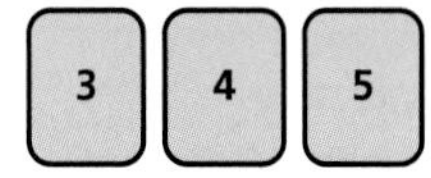

(a) Using all of these cards, Jasmine can make three-digit numbers such as 354

(i) List, in order of increasing size, all six of the three-digit numbers that Jasmine can make.

(ii) Round each of the numbers in part (i) to the nearest ten.

(iii) How many of the numbers you made in part (i) are divisible by 3?

(iv) Which of the numbers you made in part (i) are multiples of 5?

(v) Write down the number that is a hundred times the size of the largest number you made in part (i).

(vi) Write down the number that you get if you divide the smallest number in part (i) by ten.

(b) Using pairs of the same cards, Jasmine can make three proper fractions: $\frac{3}{4}$, $\frac{3}{5}$ and $\frac{4}{5}$

Which is **(i)** the smallest

(ii) the largest of these fractions?

Hint: it may help to change each fraction to a percentage.

5 Jasmine also has a set of 16 Roman number cards.

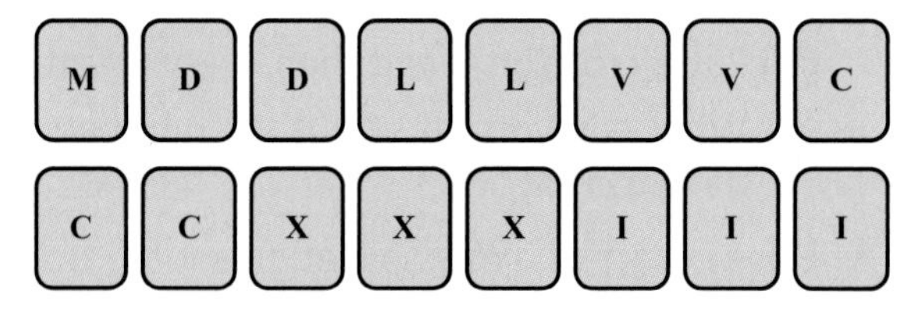

She has one M, two each of the cards D, L and V and three each of the cards C, X and I.

Jasmine says that, using these 16 cards, she could make any number up to 1999

(a) Show how Jasmine could make the number 999 in Roman numerals.

(b) How many ordinary number cards (0, 1, 2, 3, 4, 5, 6, 7, 8, 9) would Jasmine need in order to make any number up to 1999?

2 Number (2)

2.1 Properties of numbers

Prime factorisation

Every number has its own unique 'recipe' of **prime factor** 'ingredients'. There is only one way to write a given number as a product of its prime factors in order, and any given collection of prime factors will always give the same number when multiplied together. Prime factors are the 'atoms' of whole number 'molecules'.

Of course, you cannot go much further unless you know what the prime numbers are. Check the prime number table in Section 1.1, or at least remember that the first five primes are 2, 3, 5, 7 and 11

Procedure

- If you can, divide by 2 and keep dividing by 2 as many times as possible.
- Then divide by 3 as many times as possible.
- Then divide by 5, 7, 11, and so on in turn, each as many times as possible until you reach 1
- Write the original number as a product of all the primes that went into it, in order.

Examples

120, 210, 243, 225, 144

2	120
2	60
2	30
3	15
5	5
	1

$120 = 2^3 \times 3 \times 5$

2	210
3	105
5	35
7	7
	1

$210 = 2 \times 3 \times 5 \times 7$

3	243
3	81
3	27
3	9
3	3
	1

$243 = 3^5$

3	225
3	75
5	25
5	5
	1

$225 = 3^2 \times 5^2$

2	144
2	72
2	36
2	18
3	9
3	3
	1

$144 = 2^4 \times 3^2$

These last two examples (225 and 144) illustrate an important point: if the prime factors all appear an even number of times, then the original number was a square number and vice versa.

HCF and LCM

HCF

The **highest common factor (HCF)** of two or more numbers is the largest factor of the numbers, or the biggest number that divides exactly into them. For example, 15 is the biggest number that divides exactly into both 30 and 45, so we say that 15 is the HCF of 30 and 45.

It helps to consider the numbers involved as products of prime factors.

Example

18 is written as 2 × 3 × 3 and 42 is written as 2 × 3 × 7

Highlight the numbers that are common to both lists:

18 = 2 × 3 × 3   42 = 2 × 3 × 7

This highlighting shows us that the HCF of 18 and 42 is 6 (= 2 × 3).

LCM

The **lowest common multiple (LCM)** of two or more numbers is the smallest multiple of the numbers. The LCM is always a multiple of the HCF.

Example

Let us look again at 18 and 42

18 = 2 × 3 × 3 (**HCF** × 3)   42 = 2 × 3 × 7 (**HCF** × 7)

Therefore, the LCM of 18 and 42 is **HCF** × 3 × 7; **6** × 3 × 7 = 126

Addition and subtraction with negative numbers

There are four 'rules' to guide you through the process of adding and subtracting with positive and negative numbers:

- adding a *positive* is just ordinary adding
- adding a *negative* is equivalent to subtracting
- subtracting a *positive* is just ordinary subtracting
- subtracting a *negative* is equivalent to adding.

Examples

12 + ⁻3	is the same as	12 − 3 = 9
12 − ⁻3	is the same as	12 + 3 = 15
⁻5 + ⁻4	is the same as	⁻5 − 4 = ⁻9
⁻5 − ⁻4	is the same as	⁻5 + 4 = ⁻1

Multiplication and division with negative numbers

If you know your tables well, then this section will not be too hard at all.

Multiplication with negative numbers

We know that when we multiply two positive numbers we get a positive answer, so it would seem likely that if we multiplied a positive by a negative number, the answer would change to negative too. And that is what happens.

$7 \times 4 = 28$

$7 \times {}^-4 = {}^-28$

The same is true if we multiply a negative number by a positive number.

${}^-3 \times 8 = {}^-24$

Predictably, perhaps, multiplying a negative number by another negative number makes the answer positive.

${}^-5 \times {}^-9 = 45$

Division with negative numbers

Division follows very similar rules to multiplication. A positive number divided by a negative number, or vice versa, gives a negative answer, but one negative number divided by another makes the answer positive.

$72 \div {}^-8 = {}^-9$

${}^-56 \div 7 = {}^-8$

${}^-16 \div {}^-2 = 8$

Summary

We can put all this information into a table that works for both multiplication and division.

Multiply or divide this → / by this ↓	Positive	Negative
Positive	+ positive	– negative
Negative	– negative	+ positive

Revision tip

When multiplying or dividing negative and/or positive numbers, just ask yourself 'Are the signs the same?' Then the reply is 'Yes, I am *positive*!' or 'No, they are *not* (negative)'.

Exam-style questions

Try these questions for yourself. The answers are given near the back of the book.

Some of the questions involve ideas met in earlier work that may not be covered by the notes in this chapter.

2.1 Write down:

(a) the sum of 13 and 29 (1)

(b) the difference between 7 and 23 (1)

(c) the product of 8 and 9 (1)

(d) the remainder when 11 is divided by 4 (1)

(e) the square of 8 (1)

(f) the square root of 16 (1)

(g) the cube of 4 (1)

(h) the cube root of 8 (1)

(i) 3 to the power of 4 (1)

2.2 Look at these numbers.

1 3 15 18 21 39 56 60 81

From the list, write down:

(a) a multiple of 13 (1)

(b) a factor of 12 (1)

(c) the product of 8 and 7 (1)

(d) the square root of 9 (1)

(e) a cube number (1)

(f) a prime number. (1)

You may use a number more than once if you wish.

2.3 **(a)** List the prime numbers between 30 and 50 (3)

(b) Which number can be written as the product of primes $2^3 \times 3 \times 5$? (1)

(c) Write 2790 as a product of primes, using index notation. (3)

2.4 **(a)** Write these numbers in order of increasing size: (2)

$^{-}2$ 3 0 $^{-}3$ 5 $^{-}1$ 7

(b) Find the value of each of these:

(i) $^{-}3 + 5$ (1)

(ii) $1 + {}^{-}4$ (1)

(iii) $^{-}2 - 1$ (1)

(iv) $3 - {}^{-}4$ (1)

(c) Find the value of each of these:

(i) $^{-}2 \times 5$ (1)

(ii) $5 \times {}^{-}4$ (1)

(iii) $^{-}3 \times {}^{-}4$ (1)

(iv) $^{-}4 \div 2$ (1)

(v) $2 \div {}^{-}4$ (1)

(vi) $^{-}6 \div {}^{-}2$ (1)

2.2 Place value and ordering

Multiplying and dividing by 10, 100 and 1000

First remind yourself of the basics in Section 1.2 of Chapter 1.

Rule 1 Multiplying by 10 moves the decimal point right (or the digits left).

Rule 2 Dividing by 10 moves the decimal point left (or the digits right).

×	100	is	×	10	done twice
×	1000	is	×	10	done three times
÷	100	is	÷	10	done twice
÷	1000	is	÷	10	done three times

Examples

2.7	÷	10	=	0.27	1.23	×	10	=	12.3
2.7	÷	100	=	0.027	1.23	×	100	=	123
2.7	÷	1000	=	0.0027	1.23	×	1000	=	1230
0.05	×	10	=	0.5	98.7	÷	10	=	9.87
0.05	×	100	=	5	98.7	÷	100	=	0.987
0.05	×	1000	=	50	98.7	÷	1000	=	0.0987

Revision tip

If you are unsure where the decimal point goes in your answer, ask yourself if you have made the original number bigger (× 10) or smaller (÷ 10). It should then be easy! Practise these questions on your calculator by guessing the answer before you press the = button.

Ordering and rounding decimals

This section reviews the skills that will help you to read and understand decimal numbers.

Ordering decimals

Revision tip

This is made much easier by remembering that you can always write zeros at the right-hand end of any decimal number without changing its value.

0.5 is the same as 0.50 and 0.500 and 0.5000 and ...

Example

Sort these numbers into increasing order of size:

2.71 0.721 0.17 0.7 1.271 1.72

Answer

Ignore the decimal parts of each number initially and put them in whole number order.

0.721 **0**.17 **0**.7 **1**.271 **1**.72 **2**.71

Make all the decimal parts the same length by writing extra zeros where necessary.

0.721 0.17**0** 0.7**00** 1.271 1.72**0** 2.71**0**

Sort the decimal parts without disturbing the whole number order.

0.170 0.700 0.721 1.271 1.720 2.710

Remove the extra zeros.

0.17 0.7 0.721 1.271 1.72 2.71

With practice you will develop your own short cuts to this procedure, until you can do the entire sort in your head.

Rounding to two decimal places

Draw a vertical line after the second decimal place to separate the second and third positions. If the number after the line is 5 or more, then round *up*. If the number after the line is 4 or less, then round *down*.

Round *up*: add 1 to the digit *before* the vertical line. Remove all digits after the line.

Round *down*: remove all digits *after* the vertical line. Do not do anything else!

Examples

Round the numbers 3.2471 and 2.4713 to two decimal places.

3.24|71 ≈ 3.25 The 7 says we round *up*.

2.47|13 ≈ 2.47 The 1 says we round *down*.

Similarly:

2.9**5** ≈ 3.0 to 1 d.p. The 5 says round up 2.9 to 3.0

2.7181**8** ≈ 2.7182 to 4 d.p. The 8 says round up 2.7181 to 2.7182

Exam-style questions

Try these questions for yourself. The answers are given near the back of the book.

Some of the questions involve ideas met in earlier work that may not be covered by the notes in this chapter.

2.5 **(a)** Find the value of each of these:

(i) 5.07 × 10 (1)

(ii) 5.07 ÷ 100 (1)

(iii) 0.507 × 1000 (1)

(b) Use the fact that 8.2 × 13.5 = 110.7 to write down the value of each of these:

(i) 82 × 1.35 (1)

(ii) 0.82 × 13.5 (1)

(iii) 820 × 0.00135 (2)

(iv) 11070 ÷ 13.5 (2)

2.6 **(a)** Write these in order of increasing size: (2)

2.13 3.21 3.12 2.31 1.32

(b) Use the correct symbol (<, >, ≤ or ≥) to make each of these statements true:

(i) 0.023 ______ 0.203 (1)

(ii) $\frac{3}{4}$ ______ $\frac{1}{2}$ (1)

(iii) x is a number greater than or equal to 4, so x ______ 4 (1)

(iv) y is a number less than or equal to ⁻3, so y ______ ⁻3 (1)

2.3 Estimation and approximation

Significant figures

In many situations, particularly when interpreting a full calculator display, take care that the answer quoted is given to an appropriate level of accuracy. What is appropriate? Well, usually the figures in the question indicate the sort of accuracy that is required.

For example, a question asking for the circumference of a pond 4.5 m in diameter would be expecting you to write 14.1 m as your final answer, not 14.137 166 941 m. However, it is highly recommended that you show the full calculator display as part of your written working in an exam, to show what it was before you rounded it.

Remind yourself about rounding decimals by turning back to Section 2.2.

Significant figures in whole numbers
All the figures in a whole number are significant, except for zeros at the end. Thus 4500 is written to two significant figures, but 4000 is only to one significant figure. Similar ideas for rounding decimals apply when rounding whole numbers.

Examples

Round the number 23 456 to

(i) three significant figures — 2 3 4|5 6
The 5 says round *up* — 2 3 5 0 0

(ii) two significant figures — 2 3|4 5 6
The 4 says round *down* — 2 3 0 0 0

Significant figures in decimals
All the figures in a decimal are significant, except for zeros at the *beginning*. Thus 0.52 and 0.000 052 are both to two significant figures, but 0.005 200 52 is to six significant figures. (You must include the zeros in the middle.) Similarly, 0.052 000 is to *five* significant figures, since all the zeros at the end are significant.

Example

My calculator display gives the answer to a problem as 47 176.455 55

Write this to three significant figures — 4 7 1|7 6 . 4 5 5 5 5

Round *up* — 4 7 2 0 0

Notice that, as with all rounding situations, the result is the same order of magnitude (rough size) as the original number. Here both numbers (47 176.455 55 and 47 200) are roughly 47 000 (to two significant figures). Do not make the common mistake of leaving out any vital zeros. The answer is *not* 472, but 47 200!

Approximation and estimation

Sometimes we want an approximate idea of the answer to a calculation, perhaps because we are in a hurry or because we want to check that the answer we have already worked out is about right. Whatever the reason, we can find a rough answer by rounding each number to one significant figure and calculating the result, either on paper or using a calculator.

Example

$$\frac{(417.6 \times 2.91)}{(50.8 + 9.73)} \approx \frac{(400 \times 3)}{(50 + 10)} = \frac{1200}{60} = 20$$

This compares well with the exact answer which is . . . . Well, I am sure you can work it out for yourself!

Exam-style questions

Try these questions for yourself. The answers are given near the back of the book.

Some of the questions involve ideas met in earlier work that may not be covered by the notes in this chapter.

2.7 **(a)** Round:

  (i) 545 to the nearest hundred (1)

  (ii) 395 to the nearest ten. (1)

(b) Write:

  (i) 81.49 to one decimal place (1 d.p.) (1)

  (ii) 81.49 to three significant figures (3 s.f.) (1)

2.8 Estimate, to one significant figure:

(a) 5.95×7.02 (1)

(b) $81.3 \div 8.98$ (1)

2.4 Fractions, decimals, percentages and ratio

Calculating A as a fraction of B

This method is very easy!

Example

A gardener planted 30 seeds and 12 failed to germinate. What is this as a fraction?

What is 12 as a fraction of 30? Simple! $\frac{12}{30}$ That is it!

Now we can simplify this fraction in order to write it in its lowest terms:

$$\frac{12}{30} \overset{\div 2}{=} \frac{6}{15} \overset{\div 3}{=} \frac{2}{5}$$

In general, to write A as a fraction of B, just write it as $\frac{A}{B}$ and simplify as we did earlier.

Fractions as percentages

You want it as a percentage? Just make the fraction out of 100

$$\frac{2}{5} \overset{\times 2}{=} \frac{4}{10} \overset{\times 10}{=} \frac{40}{100} = 40\%$$

Remember that the *original* value or amount is the one that we call 100% when working out profits and discounts, or increases and decreases.

Example

A dress is reduced in a sale from £18 to £14.40

What is this as a percentage?

A reduction from £18 to £14.40 is a reduction of £3.60 on the original amount of £18

As a percentage of £18 this is calculated as here:

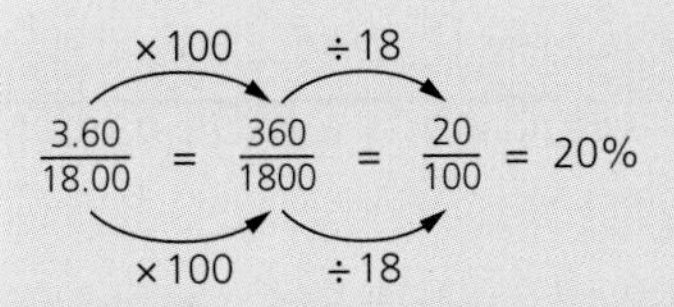

Calculating a fraction or percentage of a quantity

Remember that a fraction is simply a division into equal portions. To find a fraction of something, you simply have to divide the whole into the correct number of portions, and then take the number of portions that are required.

Example (using fractions)

Find $\frac{3}{4}$ of £1.80

The first thing to notice is that we want to divide the money into four equal portions (quarters); then we want three of these portions.

To find $\frac{1}{4}$ of £1.80 we simply divide it by 4. Always **d**ivide by the **d**enominator!

$4\overline{)1.80}$ = 0.45 or $4\overline{)180}$ = 45

so $\frac{1}{4}$ of £1.80 is 45p.

Now, $\frac{3}{4}$ is 3 lots of $\frac{1}{4}$

therefore $\frac{3}{4}$ of £1.80 is £1.35 (3 × 45p).

Revision tip

Divide by the **D**enominator, then **T**imes by the **T**op.

An easy trick with percentages is to base all your working around the calculation of 10%, which is simple to find since it is just one-tenth. 5% is then half of it, 20% is double it and so on.

Example (using percentages)

Find 35% of 140 kg.

	10% of 140 kg is 14 kg (one tenth)
Therefore:	5% of 140 kg is 7 kg (half of 10%)
	30% of 140 kg is 42 kg (3 × 10%)
Therefore:	35% of 140 kg is 49 kg (5% + 30%)

Another trick is to use percentages of 100 which, of course, do not even need to be worked out! If you want to find 36% of 50 cm, just use the fact that 36% of 100 cm is 36 cm, and then halve it to get 18 cm. Similarly, to find 17% of 200, just say '17% of 100 is 17', and then it is clear that 17% of 200 must be 34.

Fraction, decimal and percentage equivalents

Here is a simple little diagram that has helped many pupils through the fraction/percentage/decimal maze. I hope it helps you too!

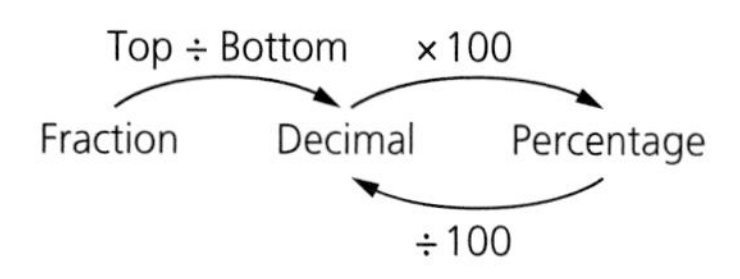

Converting a decimal to a fraction is a little harder, but can be done in a few steps:

- Count the digits after the decimal point. For one digit think 'ten', for two digits think 'hundred' and so on, with each extra digit causing you to think of an extra zero.
- Write the digits after the decimal point as the numerator and the number in your head as the denominator.
- Now just cancel down this fraction in the normal way.

Examples

$0.25 = \frac{25}{100} = \frac{5}{20} = \frac{1}{4}$

$0.175 = \frac{175}{1000} = \frac{35}{200} = \frac{7}{40}$

Writing a fraction as a decimal and as a percentage

Write the fraction $\frac{3}{4}$ as a decimal and as a percentage.

$3 \div 4 = 0.75$ or $4\overline{)3.00}$ with 0.75 on top if you are doing it by hand.

Now to write 0.75 as a percentage, we simply multiply by 100:

$0.75 \times 100 = 75\%$

Ordering fractions, decimals and percentages

Place these in increasing order of size:

0.9 78% $\frac{4}{5}$ $\frac{7}{8}$

Method

Write as decimals:	0.9	0.78	0.8	0.875
Sort the decimals:	0.78	0.8	0.875	0.9
Rewrite as they were:	78%	$\frac{4}{5}$	$\frac{7}{8}$	0.9

Addition, subtraction, multiplication and division of fractions

Fractions are like other numbers. They can be added, subtracted, multiplied or divided. It is not always easy though!

A reminder of the basics

- You cannot add or subtract two fractions with different denominators. You must make the denominators the same first.
- To make two denominators the same, find the smallest number into which both denominators divide (the lowest common multiple or lowest common denominator) and use equivalent fractions to scale up the two fractions accordingly.

- When adding mixed numbers, work out the whole number part first and then add the fractions. When subtracting, multiplying or dividing fractions, convert mixed numbers into improper 'top heavy' fractions before proceeding. This can make subtractions easier; for multiplication and division it is vital.

Level 1
While Level 1 and Level 2 cover the same syllabus, the questions set for Level 1 will generally be more straightforward. For example, ISEB will test the addition and subtraction of fractions only with the same denominator.

Level 3
Level 3 requires the addition and subtraction of both fractions with different denominators and mixed fractions. It also requires the multiplication and division of both proper and mixed fractions.

This table shows the sorts of calculations you can expect to meet at each level.

	All levels	Level 3 only
+	$\frac{3}{7}+\frac{2}{7}=\frac{5}{7}$, $1\frac{1}{2}+\frac{2}{3}=2\frac{1}{6}$	
−	$\frac{8}{11}-\frac{5}{11}=\frac{3}{11}$, $1\frac{1}{3}-\frac{3}{4}=\frac{7}{12}$	
×	$\frac{3}{4}\times\frac{5}{7}=\frac{15}{28}$, $\frac{3}{5}\times\frac{2}{3}=\frac{2}{5}$	$1\frac{1}{3}\times\frac{3}{4}=1$
÷	$\frac{4}{5}\div\frac{2}{5}=2$, $\frac{3}{4}\div\frac{2}{3}=1\frac{1}{8}$	$2\frac{1}{2}\div1\frac{3}{4}=1\frac{3}{7}$

Example 1 (addition)

$3\frac{3}{4}+2\frac{4}{5}$ is the same as $5+\frac{3}{4}+\frac{4}{5}$

Now, $\frac{3}{4}+\frac{4}{5}=\frac{15}{20}+\frac{16}{20}=\frac{31}{20}$ or $1\frac{11}{20}$

So $3\frac{3}{4}+2\frac{4}{5}=5+1\frac{11}{20}=6\frac{11}{20}$

Example 2 (subtraction)

$3\frac{3}{4}-2\frac{4}{5}$ First change to improper fractions.

$3\frac{3}{4}-2\frac{4}{5}=\frac{15}{4}-\frac{14}{5}$

Remember that when you write a mixed number as an improper fraction the denominator stays the same.

To find the new numerator, use the word formula:

new numerator = whole number × denominator + old numerator

Then proceed as usual.

$\frac{15}{4}-\frac{14}{5}=\frac{75}{20}-\frac{56}{20}=\frac{19}{20}$

Example 3 (multiplication)

$3\frac{3}{4}\times2\frac{4}{5}=\frac{15}{4}\times\frac{14}{5}$

$\frac{3}{4}\times\frac{14}{1}$ (cancel 5s)

$\frac{3}{2}\times\frac{7}{1}=\frac{21}{2}$ or $10\frac{1}{2}$ (cancel 2s)

Example 4 (division)

$3\frac{3}{4} \div 2\frac{4}{5} = \frac{15}{4} \div \frac{14}{5}$

invert the second fraction

$\frac{15}{4} \times \frac{5}{14} = \frac{75}{56}$ or $1\frac{19}{56}$

Calculating with ratios

It often helps to think of ratios as parts, and to begin by seeing how many parts are involved altogether. That way they become fractions.

Example 1

Share £42 in the ratio 2:5

2 parts + 5 parts = 7 parts altogether, so each part is $\frac{1}{7}$ of £42, in other words £6

Answer

2 parts = 2 × £6 and 5 parts = 5 × £6

So £42 shared in the ratio 2:5 is £12:£30

Example 2

A class contains boys and girls in the ratio 3:2

If there are 12 girls, how many boys are there?

First be clear that because of the order of the words, the 3 in the ratio goes with *boys* and the 2 goes with *girls*.

We are told that the 2 in the ratio represents 12 girls, so each part of the ratio represents 6 children.

Answer

The number of boys is 3 (from the ratio) multiplied by 6 (the number of children per part). So there are 18 boys in the class.

Adapting a recipe

Sometimes you have to change the quantities in a recipe so that it serves a different number of people. Ratios to the rescue!

Example 3

These quantities are used to make an apple pie that serves four people.

200 g shortcrust pastry

480 g cooking apples

120 g brown sugar

4 cloves

Adapt this recipe to serve five people.

250 g shortcrust pastry

600 g cooking apples

150 g brown sugar

5 cloves

We first calculate one person's share (dividing by 4) and then multiply this by 5, as has been done here.

This method is called the unitary method, and can be used for either increases or decreases.

Exam-style questions

Try these questions for yourself. The answers are given near the back of the book.

Some of the questions involve ideas met in earlier work that may not be covered by the notes in this chapter.

2.9 **(a)** Write the fraction representing 5 parts of a whole containing 8 parts. (1)

(b) What fraction of this rectangle has been shaded? (1)

■	■	■	■					
■	■	■	■					

(c) Using the words denominator and numerator, name the parts of the fraction $\frac{2}{3}$ (1)

2.10 **(a)** Change the fraction $\frac{3}{5}$ into:

(i) an equivalent fraction with denominator 10 (1)

(ii) a decimal (decimal fraction) (1)

(iii) a percentage. (1)

(b) Write the fraction $\frac{18}{30}$ in its simplest form (lowest terms). (1)

2.11 **(a)** **(i)** Write the improper fraction $\frac{17}{5}$ as a mixed number. (1)

(ii) Write the mixed number $2\frac{2}{5}$ as an improper fraction. (1)

(b) Write the mixed number $1\frac{3}{4}$ as:

(i) a decimal (1)

(ii) a percentage. (1)

2.12 Calculate:

(a) $\frac{3}{13} + \frac{7}{13}$ (1)

(b) $\frac{1}{2} + \frac{2}{3}$ (1)

(c) $1\frac{1}{3} + \frac{3}{4}$ (2)

(d) $\frac{7}{8} - \frac{3}{8}$ (1)

(e) $\frac{2}{3} - \frac{1}{2}$ (1)

(f) $1\frac{1}{3} - \frac{3}{4}$ (2)

2.13 Calculate:

(a) $\frac{3}{4} \times \frac{2}{3}$ (1)

(b) $1\frac{1}{4} \times \frac{2}{3}$ (2)

(c) $\frac{1}{2} \div \frac{2}{3}$ (2)

Level 3 ■ **(d)** $1\frac{1}{3} \div 2\frac{3}{4}$ (3)

2.14 **(a)** There are 11 pupils in a school team. Six are boys. Write the ratio:

(i) boys to the total (1)

(ii) boys to girls (1)

(iii) girls to boys. (1)

(b) Write the ratio 8 : 12 in its simplest form. (1)

2.15 To make six chocolate crunchies, these quantities are used.

Cornflakes	150 g
Chocolate	125 g
Honey	10 ml
Sultanas	25 g

Rewrite the recipe to make 24 chocolate crunchies. (4)

★ Make sure you know

- ★ All of the material in Chapter 1
- ★ What prime factors are, and how to use them to express a number as a product of primes, using index notation
- ★ How to order, add and subtract negative numbers in the abstract
- ★ How to use your understanding of place value to multiply and divide whole numbers and decimals by 10, 100 and 100
- ★ When making estimates, that you should round to one significant figure and multiply and divide mentally
- ★ How to reduce a fraction to its simplest form by cancelling common factors and solve simple problems involving ratio and direct proportion
- ★ How to calculate fractional or percentage parts of quantities and measurements, using a calculator where appropriate
- ★ That you should be aware of which number to consider as 100 per cent, or a whole, in problems involving comparisons, and use this to evaluate one number as a fraction or percentage of another
- ★ The equivalences between fractions, decimals and percentages, and how to calculate using ratios in appropriate situations
- ★ How to add and subtract fractions by writing them with a common denominator
- ★ How to multiply and divide fractions, including easy mixed numbers
- ★ How to use proportional changes, calculating the result of any proportional change using only multiplicative methods
- ★ That you should be using the glossary at the back of the book for looking up definitions of key words

Test yourself ✓

Before moving on to the next chapter, make sure you can answer these questions. The answers are near the back of the book.

1 Complete this jumbled multiplication table as quickly as you can. Record your time in minutes and seconds. (A time of less than 2 minutes is quite good!)

×	4	9	2	5	8	6	1	3	7
3									
9									
8									
2									
6									
5									
4									
7									
1									

2 Look at these numbers.

4 5 8 16 19 27 30 32

From the list, choose a *different* answer for each of these:

(a) a prime number
(b) the square root of 16
(c) a factor of 15
(d) a multiple of 4
(e) a square number
(f) a cube number.

3 Look at these eight directed number cards.

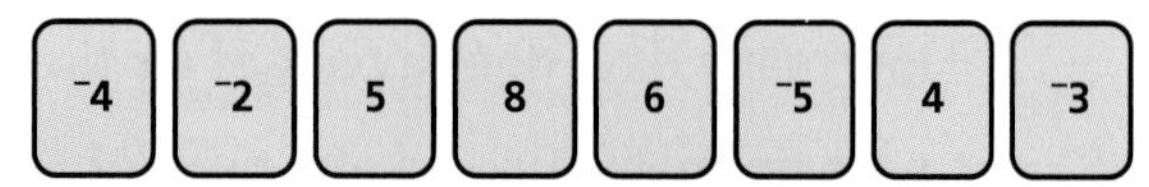

(a) What is the sum of all eight cards?
(b) Choose two cards that have a sum of zero.
(c) Which two cards give the lowest product?
(d) Using all eight directed number cards once only, fill in the gaps in these calculations.
  (i) _____ + _____ = 1
  (ii) _____ − _____ = 2
  (iii) _____ × _____ = 15
  (iv) _____ ÷ _____ = $^{-}4$

4 **(a)** Express the number 630 as the product of prime factors, using indices.
(b) Evaluate $2^3 \times 3^2 \times 5 \times 7^2$

5 **(a)** Write these numbers in order of size, starting with the smallest:

6.606 6.6 6 6.066 6.66

(b) What is the difference between the largest and the smallest of the numbers in part (a)?

6 By first writing each number correct to one significant figure, estimate the value of these:

(a) $\dfrac{59.3 \times 30.4}{5.75}$

(b) $\dfrac{41.7 \times 3.2}{0.63 \times 49.4}$

7 Complete this table to show rows of equivalent fractions, decimals and percentages:

Fraction	Decimal	Percentage
$\frac{7}{20}$	0.35	35%
$\frac{1}{4}$		
	1.2	120%
		15%

8 **(a)** Express 26% as a fraction in its lowest terms.
(b) Write $\frac{13}{20}$ as a decimal.
(c) Calculate $\frac{2}{3}$ of £4.80
(d) Express 0.54 as a fraction in its lowest terms.
(e) Find 15% of £40

9 (a) Last year in Roley there were 50 reported cases of 'flu. This year the number dropped by 12%.

How many cases have been reported this year?

(b) A farmer digs up 120 kg of potatoes and he finds that 20% are rotten. Of the remaining healthy potatoes, 12.5% are too small to sell. What is the mass of healthy potatoes large enough to sell?

(c) At a party, 15 children share three identical cakes equally. What fraction of a cake does each child receive?

(d) My cat, Tabitha, eats $\frac{3}{5}$ of a tin of cat food each day. How many days does a pack of six tins last Tabitha?

10 Shirts are normally marked at a price of £45 each.

In a sale I paid £36 for a shirt.

(a) How much money did I save?

(b) What was the percentage discount for the sale?

On a second shirt, I was offered a further 10% discount on the sale price.

(c) How much did I pay for the second shirt?

(d) What was the overall percentage reduction on the original marked price for the second shirt?

11 (a) Evaluate:

(i) $\frac{1}{4} + \frac{1}{6}$

(ii) $\frac{3}{4} - \frac{1}{6}$

(b) Evaluate:

(i) $\frac{1}{3} \times \frac{3}{4}$

(ii) $\frac{1}{3} \div \frac{3}{4}$

(c) Evaluate:

Level 3

■ **(i)** $2\frac{1}{3} \times 1\frac{3}{4}$

■ **(ii)** $2\frac{1}{3} \div 1\frac{3}{4}$

12 One sunny day, a man stood near to a tree.

The shadow of the tree was 10.8 metres long.

The man was 1.8 metres tall and his shadow was 2.7 metres long.

What was the height of the tree?

13 The sizes of the angles of a quadrilateral are in the ratio 1 : 2 : 1 : 2

(a) Calculate the size of each angle.

(b) What shape is the quadrilateral?

14 A blend of coffee is made up of three parts Brazilian and two parts Honduran by weight.

(a) How many grams of each type of coffee are there in 200 g of the blend?

The costs of Brazilian and Honduran coffee are in the ratio 3 : 5. Brazilian coffee costs £12 per kilogram.

(b) What is the cost of a kilogram of Honduran coffee?

(c) What should be the cost of a jar containing 200 g of the blend? Give your answer correct to the nearest 10 pence.

3 Calculations (1)

3.1 Number operations

The number line can be very useful in understanding the number operations.

Addition

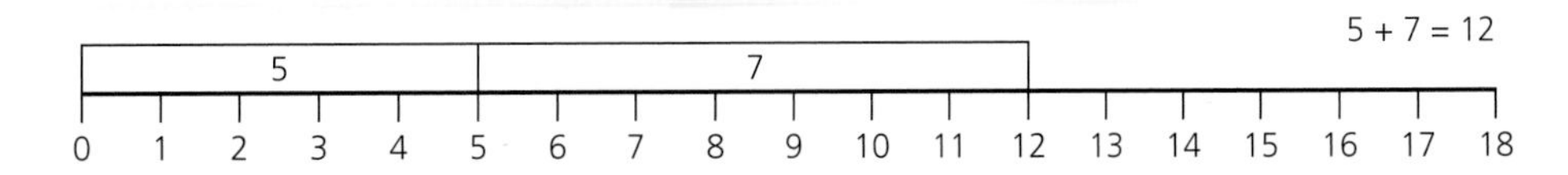

Subtraction

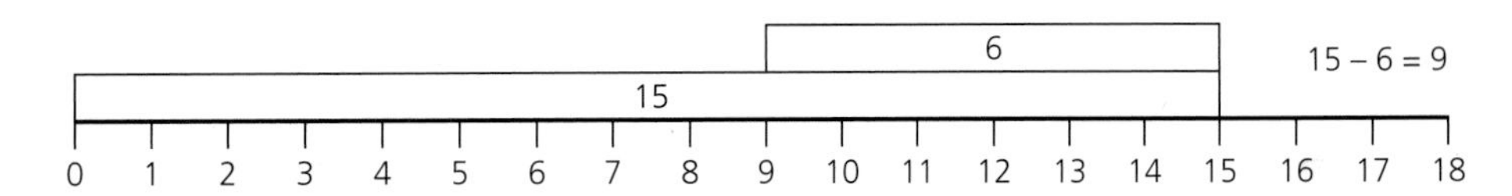

Multiplication

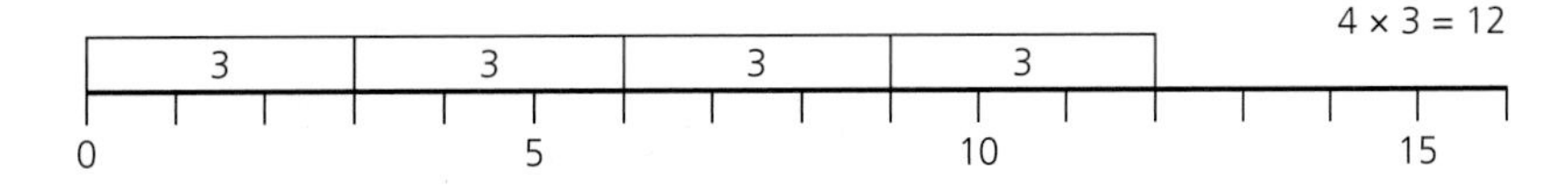

Division – exact

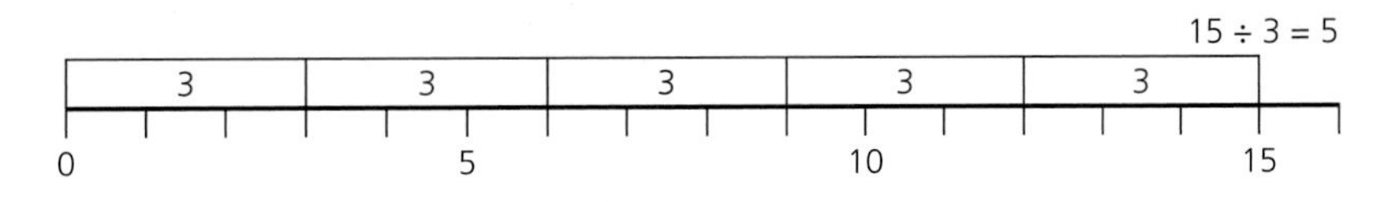

Division – with a remainder

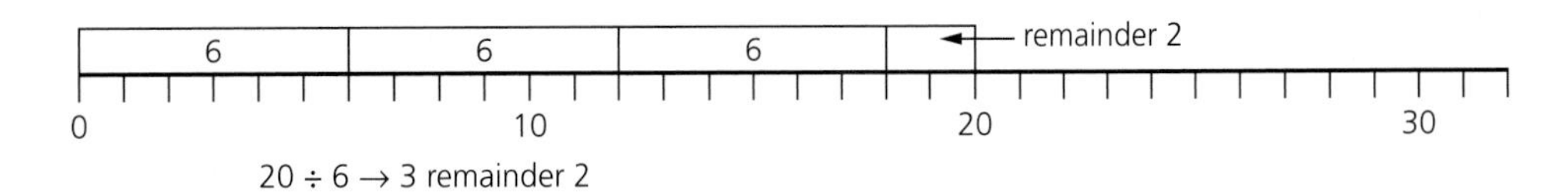

It is also useful to see what happens with square numbers and prime numbers.

Square numbers

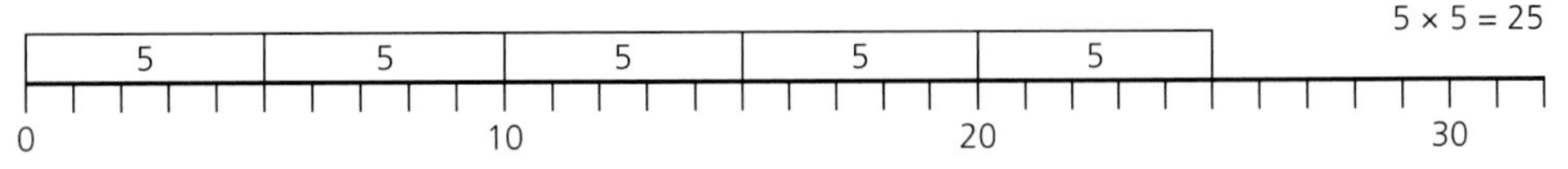

Prime numbers

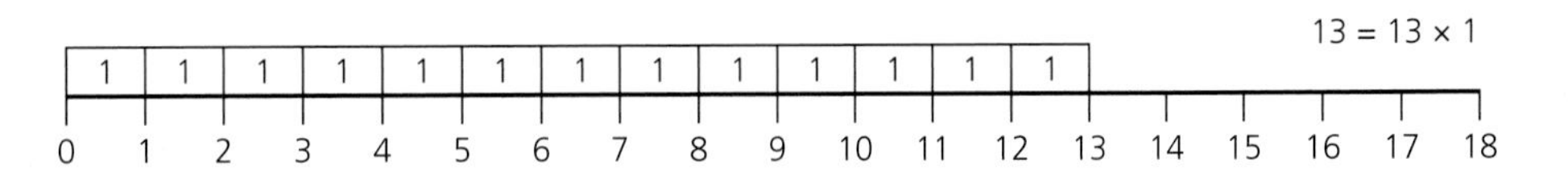

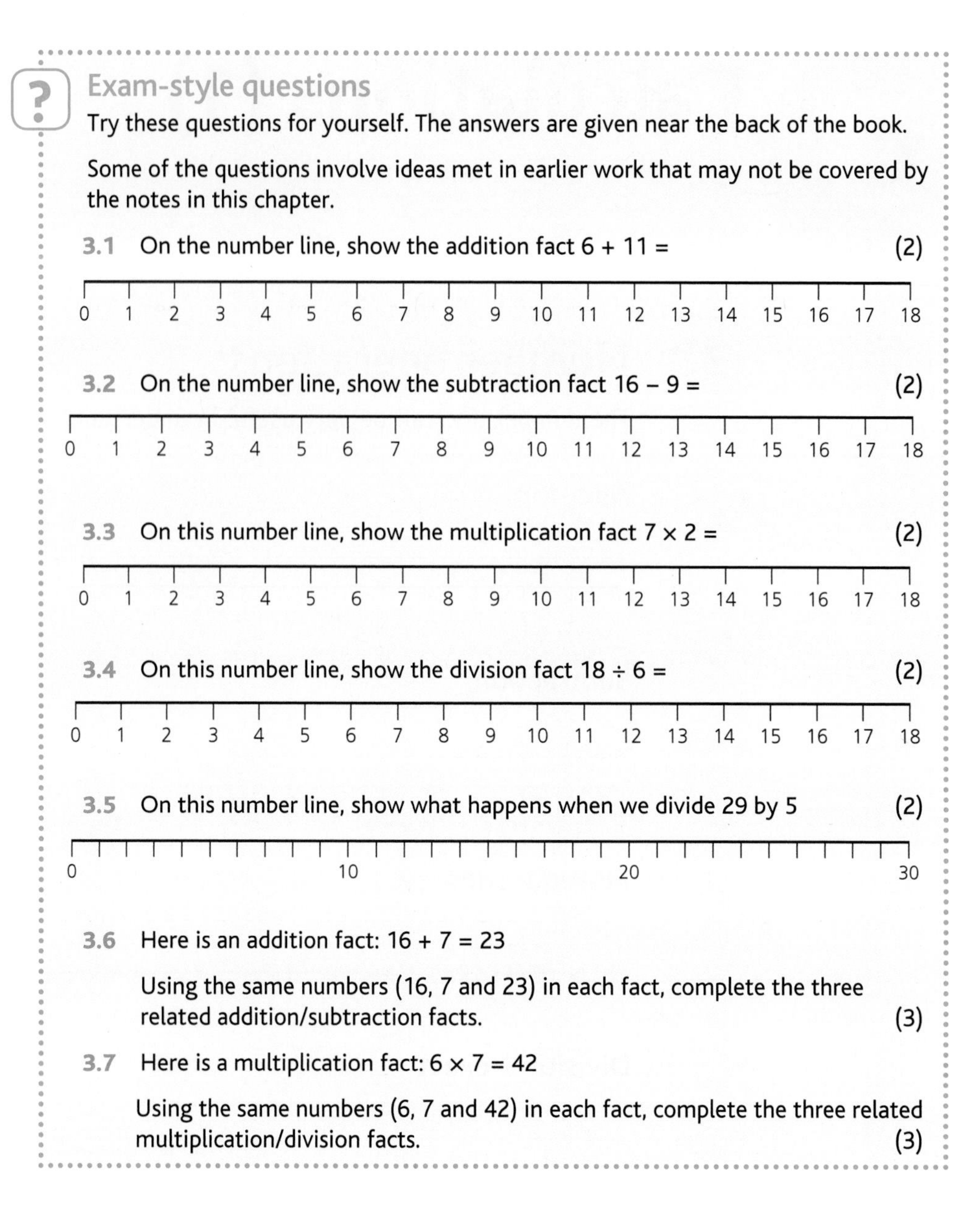

Exam-style questions

Try these questions for yourself. The answers are given near the back of the book.

Some of the questions involve ideas met in earlier work that may not be covered by the notes in this chapter.

3.1 On the number line, show the addition fact 6 + 11 = (2)

0 1 2 3 4 5 6 7 8 9 10 11 12 13 14 15 16 17 18

3.2 On the number line, show the subtraction fact 16 – 9 = (2)

0 1 2 3 4 5 6 7 8 9 10 11 12 13 14 15 16 17 18

3.3 On this number line, show the multiplication fact 7 × 2 = (2)

0 1 2 3 4 5 6 7 8 9 10 11 12 13 14 15 16 17 18

3.4 On this number line, show the division fact 18 ÷ 6 = (2)

0 1 2 3 4 5 6 7 8 9 10 11 12 13 14 15 16 17 18

3.5 On this number line, show what happens when we divide 29 by 5 (2)

0 10 20 30

3.6 Here is an addition fact: 16 + 7 = 23

Using the same numbers (16, 7 and 23) in each fact, complete the three related addition/subtraction facts. (3)

3.7 Here is a multiplication fact: 6 × 7 = 42

Using the same numbers (6, 7 and 42) in each fact, complete the three related multiplication/division facts. (3)

3.2 Mental strategies

Mental strategies

It is an excellent idea to have a range of strategies for speeding up calculations in your head.

Change the question!

When working out 7 × £2.99 it is much easier to change it to 7 × £3 and then subtract those seven extra pennies at the end to get £20.93

Subtraction and addition in stages

If you want to subtract something awkward like 54, then it is usually easier to subtract in two or more stages: first subtract the 50, then subtract the 4

Adding in stages is surprisingly similar …

Subtraction and addition in partnership

It is often a really good idea to use a combination of adding and subtracting to make the calculation easier. For example, when subtracting 17, it is probably easier to subtract 20 then add on 3 to the result.

The famous five

Multiplying by 5 is equivalent to multiplying by 10 and then halving the answer. Dividing by 5 is equivalent to dividing by 10 and doubling the answer. You need to try this out to see just how quick this is!

Nines

$10 - 1 = 9$

OK, so you knew that (I hope!) but it does give us a very handy way of multiplying by 9: just multiply by 10 then subtract the original number.

Elevens

$10 + 1 = 11$

You can probably see where this is going ... To multiply by 11, just multiply by 10 then add the original number.

Last digit delights

The last digit in any product is the product of the last digits in the two numbers being multiplied. So, for example, we know that 24×312 must end in an 8 since 8 is the result of 4×2 in the original question.

Playing rough

Always get a rough answer before you work anything out to give you a quick check that you have not messed it up completely!

Exam-style questions

Try these questions for yourself. The answers are given near the back of the book.

Some of the questions involve ideas met in earlier work that may not be covered by the notes in this chapter.

Note: For all questions in this section you should do no written working, but simply write the answers. Do all the calculations in your head. Check that each answer is sensible.

3.8 Fill in the missing numbers.

(a) $4 \times 19 =$ (1)

(b) $7 \times 31 =$ (1)

(c) $29 \times 5 =$ (1)

(d) $11 \times 17 =$ (1)

3.9 There are 254 children on a school's register. One day 59 were absent. How many children were at school that day? (2)

3.10 There are 37 people on a bus. At the next stop 19 of them get off and 11 get on. How many people are now on the bus? (3)

3.11 What is the cost of 11 chocolate bars costing 43 pence each? (2)

3.12 What is 9.5×4? (2)

3.13 Sarah left home at 08:45 and cycled for 35 minutes to reach her friend's house. At what time did she arrive? (2)

3.14 What is the smallest number, greater than 200, which divides exactly by 3? (2)

3.15 A known division fact is $32 \div 8 = 4$

Use this fact to help you to find the answers to these:

(a) $320 \div 8 =$ ______ (1)

(b) $32 \div 80 =$ ______ (2)

(c) $3200 \div 80 =$ ______ (1)

(d) $32 \div 0.8 =$ ______ (2)

(e) $3.2 \div 8 =$ ______ (2)

3.16 (a) Anne used her calculator to share £52 between eight people.

This was her calculator display: 6.5
How much money did each person receive? (1)

(b) Barney's calculator shows this result: 30
Which number had he multiplied by 4? (2)

(c) Carrie's calculator shows this display: −4
Which number did Carrie subtract from 7? (2)

3.3 Written methods

Addition and subtraction of decimals to two places

If you can add and subtract whole numbers reliably then this section should not be too difficult at all.

Revision tip

There is really only one thing to remember here.

Line up the decimal points! ... Or if you prefer ... Line up the units digits!

Addition

Let us try adding three numbers together, bearing this rule in mind:

Examples

U.t		U.t h		U.t h
2. 1	+	0 .2 1	+	2 .1 2

```
  U.t h
  2.1
  0.2 1
+ 2.1 2
  -----
  4.4 3   Correct!
```

Subtraction

This is very similar – just make sure that you line up those decimal points. Remember also that writing zeros at the end of a decimal does not change the number, and that putting .0 or .00 on the end of a whole number is OK too.

Examples

4.78 − 3.6	7.1 − 4.68	7 − 0.37
4.78 −3.60 1.18	(carries: 6, 10, 1) 7.10 −4.68 2.42	(carries: 6, 9, 1) 7.00 −0.37 6.63
(just like doing 478 − 360)	(just like doing 710 − 468)	(just like doing 700 − 37)

Multiplication and division by a single-digit number

The first thing to do after you have learned your tables is to learn to multiply or divide any large number by a single-digit number.

Short multiplication – multiplying by a single-digit number (e.g. 126 × 7)

The two favourite methods are the compact and partitioning methods:

Compact

$$\begin{array}{r} 126 \\ \times\ 7 \\ \hline \mathbf{882} \\ \hline {\scriptstyle 14} \end{array}$$

Partitioning

$$\begin{array}{rcl} 100 \times 7 &=& 700 \\ 20 \times 7 &=& 140 \\ 6 \times 7 &=& 42 \\ \hline && \mathbf{882} \end{array}$$

Short division

Simple short division	Division with remainder as a decimal
384 ÷ 6 = 64 064 6⟌384 (carries 3, 2) The zero above the three is important because it acts as a place holder and stops us getting a crazy answer such as 640 or 604! It is always a good idea to write it in.	384 ÷ 5 = 76.8 076.8 5⟌384.0 (carries 3, 3, 4) As before, we must write the zero above the three. We also write in two decimal points, one above the other, so that we can continue the division beyond it.
Check by multiplying: 64 × 6 384 (carries 3 2) Obviously this should take us back to our original number. And it does!	Division with remainder as a fraction $384 \div 7 = 54\frac{6}{7}$ 054 7⟌384 (carries 3, 3) remainder 6 Since we were dividing by 7, the remainder is put over 7 to make the fraction.

Long multiplication

Long multiplication means multiplying together two numbers that are too big to work with in your head.

We shall take the example 126 × 64 and work it through, explaining every stage in detail.

Procedure

1 Write the two-digit number underneath the three-digit number, lining up the units column as if it were an addition problem.

$$\begin{array}{r} 1\,2\,6 \\ \times \quad 6\,4 \\ \hline \end{array}$$

2 Multiply the three-digit number by the units digit of the two-digit number, working from right to left, with carrying where necessary.

$$\begin{array}{r} 1\,2\,6 \\ \times \quad 6\,4 \\ \hline 5\,0\,4 \\ {\scriptstyle 1\;\;2\;\;\;} \end{array}$$

3 Put a zero in the units column of the second row of working. Multiply the three-digit number by the tens digit of the two-digit number, again working from right to left, with carrying if it is needed.

$$\begin{array}{r} 1\,2\,6 \\ \times \quad 6\,4 \\ \hline 5\,0\,4 \\ 7\,5\,6\,0 \\ {\scriptstyle 1\;\;3\;\;\;\;\;\;\;} \end{array}$$

4 Add together the two rows of working to obtain the final answer.

$$\begin{array}{r} 1\,2\,6 \\ \times \quad 6\,4 \\ \hline 5\,0\,4 \\ 7\,5\,6\,0 \\ \hline 8\,0\,6\,4 \\ \hline {\scriptstyle 1\;\;\;\;\;\;\;\;\;\;\;\;} \end{array}$$

Notes: You can, if you wish, multiply by the tens digit first, but it is still a good idea to keep two separate rows of working. The reason we write a zero in the units column in step 3 is because we are really multiplying by 60, not by 6, so writing the 0 effectively does the multiplying by 10 for us.

Long division

The procedure for long division involves a cycle of four operations 'DMSB' that are repeated until the answer is obtained. Long division with bigger numbers is not necessarily harder, it just takes more cycles of DMSB until the working out is finished. These are the steps:

D Division

M Multiplication

S Subtraction

B Bring down the next digit

Revision tip

Think up a mnemonic to help you remember the sequence DMSB.

'Don't Make Silly Blunders' is the classic reminder.

Can you think of a better one?

Remember that any division involves four numbers, each with its own name.

Consider the division $17 \div 5 = 3$ r 2

17 is the **dividend**

5 is the **divisor**

3 is the **quotient** (or simply the **answer**)

2 is the **remainder**

When we do long division we start by writing out the first nine multiples of the divisor. It is helpful to write this at the right-hand side of the page for reference.

Example

Let us begin with the long division 7658 ÷ 24

First we write out the first nine multiples of the divisor, which in this case is 24.

1 × 24 = 24

2 × 24 = 48

3 × 24 = 72

4 × 24 = 96

5 × 24 = 120

6 × 24 = 144

7 × 24 = 168

8 × 24 = 192

9 × 24 = 216

The working out, in full, looks like this when complete:

```
          0  3  1  9
        ____________
2  4   | 7  6  5  8
         0
        ___
         7  6
         7  2
        ______
            4  5
            2  4
           ______
            2  1  8
            2  1  6
           _________
                  2
```

First cycle of DMSB:

D 7 ÷ 24 = 0 (we always ignore remainders) so write 0 above the 7 in the quotient (answer) line at the top.
M 0 × 24 = 0 (the previous result multiplied by the divisor) so write 0 below the 7 in the dividend.
S 7 – 0 = 7 so draw a line and write 7 below the 0 in the working.
B Bring down the 6 so that we are now dividing 24 into 76

Now we DMSB again on the 76

Second cycle of DMSB:

D 76 ÷ 24 = 3 (using our multiples of 24 chart to find the greatest multiple less than or equal to 76) so write 3 above the 6 in the quotient (answer) line at the top.
M 3 × 24 = 72 (again using our multiples of 24 chart) so write 72 below the 76 in the dividend.
S 76 – 72 = 4 so draw a line and write 4 below the 72 in the working.
B Bring down the 5 to change the 4 into 45

Now we DMSB again on the 45

Third cycle of DMSB:

D 45 ÷ 24 = 1 so write 1 above the 5 in the quotient (top) line.
M 1 × 24 = 24 so write 24 below the 45 in the working.
S 45 – 24 = 21 so draw a line and write 21 below the 24 in the working.
B Bring down the 8 to change 21 into 218

Finally do DMSB on the 218

Fourth cycle of DMSB:

D 218 ÷ 24 = 9 so write 9 above the 8 in the quotient line.
M 9 × 24 = 216 so write 216 below the 218 in the working.
S 218 − 216 = 2 so draw a line and write 2 below the 216 in the working.
B There is nothing left to bring down so we leave the 2 as the remainder.

The answer is 319 r 2. Done!

Note: If the answer is required as a decimal instead of with a whole number remainder, then simply write in a decimal point after the dividend and follow it with as many zeros as you want decimal places in the answer.

From here on you can then carry on with DMSB as before, with the only difference being that you bring down a zero at each stage.

Exam-style questions

Try these questions for yourself. The answers are given near the back of the book.

Some of the questions involve ideas met in earlier work that may not be covered by the notes in this chapter.

Note: When answering these questions, you are expected to show clearly all your working, even if you could do them in your head.

3.17 (a) Add 77 to 48 (1)
(b) Subtract 19 from 102 (2)

3.18 (a) Add 2.9 + 4.6 (1)
(b) Subtract 3.75 from 12.2 (2)

3.19 (a) Which number is 3.7 less than 10.4? (1)
(b) What must I add to 9.6 to get 14.1? (2)

3.20 (a) Multiply 18 by 7 (2)
(b) Multiply 312 by 13 (2)

3.21 (a) Divide 1205 by 5 (2)
(b) Divide 1001 by 7 (2)

3.22 (a) Divide 312 by 13 (2)
(b) Divide 117 by 45, giving your answer as a decimal. (2)

3.23 (a) Divide 1000 by 14, giving the answer with a remainder. (2)
(b) Write your answer to part (a) as a mixed fraction. (1)

3.24 (a) Divide 319 by 7, giving your answer to the nearest whole number. (2)
(b) Divide 1440 by 17, giving your answer to the nearest ten. (2)

3.25 (a) What is the total cost of six pens costing £2.75 each? (2)
(b) Share £2 between seven people as best you can.
(i) How many pence will each person receive? (2)
(ii) How much will be left over? (2)

★ Make sure you know

★ And understand fully the four operations (addition, subtraction, multiplication and division) and the associated language

★ How to use a range of mental methods of computation with the four operations when solving number problems (including mental recall of multiplication facts up to 12 × 12 and quick derivation of corresponding division facts)

★ How to check the reasonableness of your results by reference to your knowledge of the context or to the size of the numbers

★ How to interpret a calculator display

★ How to use efficient written methods of addition and subtraction and of short multiplication and division (including addition and subtraction of decimals to two places, and ordering decimals to three places)

★ How to multiply numbers up to four digits by a two-digit number using the formal written method of long multiplication

★ How to divide numbers up to four digits by a two-digit number using the formal method of long division, and interpret remainders

★ How to use all four operations with decimals to two places (where multiplication and division are always by an integer less than ten), and rounding to the nearest integer

★ How to use and understand an appropriate non-calculator method for multiplying any three-digit number by any two-digit number

★ How to use the glossary at the back of the book for definitions of key words

Test yourself ✓

Before moving on to the next chapter, make sure you can answer these questions. The answers are at the back of the book.

1 Brian has multiplied 197 by 15 and has written the answer incorrectly as 1182

(a) Suggest at least two reasons why Brian should realise straight away that his answer is incorrect.

(b) What should the answer be?

(c) Suggest what error Brian made.

2 Calculate the square of 19

3 Which number between 240 and 250 divides exactly by 7?

4 Jasmine still has her three number cards!

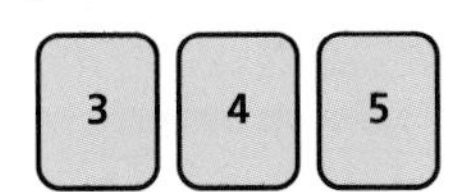

Jasmine arranges her cards into a two-digit number and a single-digit number, for example, 35 and 4

She then finds the product of these two numbers, so 35 × 4 is 140

Which similar grouping of the cards will give

(a) the smallest product

(b) the largest product?

5 Diana used her calculator while doing some shopping to keep a record of how much money she would have left.

First she entered 20, since she started with a £20 note. She subtracted the cost of each item as she put it into her shopping basket.

Just before she reached the checkout she pressed [=] and was surprised to see this display.

Assuming that she had pressed the keys correctly, what does this mean?

4 Calculations (2)

Before you start this chapter, it might be a good idea to have another look at Chapter 3, particularly the section on mental strategies. You might also like to answer questions 3.8 to 3.16 again.

4.1 Written methods

Multiplication and division of decimals to two decimal places

This is very similar to working with whole numbers – once you can do one you can do the other.

Multiplying

- Add together the *numbers* of decimal places in the two numbers being multiplied.
- Rewrite the questions without the decimal points.
- Multiply the whole numbers in the usual way.
- Count back the total number of decimal places and put the decimal point in the answer.

Example (multiplying decimals)

Calculate 2.3 × 4.76 without using a calculator.

2.3	×	4.76	
1 d.p.		2 d.p.	Total: 3 d.p.

Rewrite and calculate as a long multiplication without the decimal points.

$$\begin{array}{r} 476 \\ \times \quad 23 \\ \hline 1428 \\ 9520 \\ \hline 10948 \\ \hline \end{array}$$

Count back three places and rewrite the decimal point.

This gives the final answer **10.948**

Quick check: 2.3 × 4.76 is approximately 2 × 5 = 10 so this seems about right.

Whenever you multiply or divide, always make an estimate first to help you check your answer!

Dividing

- Write the division as a fraction, with one number above the other.
- Multiply top and bottom by 10 (if either number is to 1 d.p.) or by 100 (if either number is to 2 d.p.).
- Cancel down the fraction in the usual way to simplify it as much as possible.

Example (dividing decimals)

Calculate $3.15 \div 4.2$

First write as a fraction $\frac{3.15}{4.2} \times \frac{100}{100} = \frac{315}{420}$

Notice that we multiply top and bottom by 100 because the first number has two decimal places.

Now cancel down to produce the simplified answer.

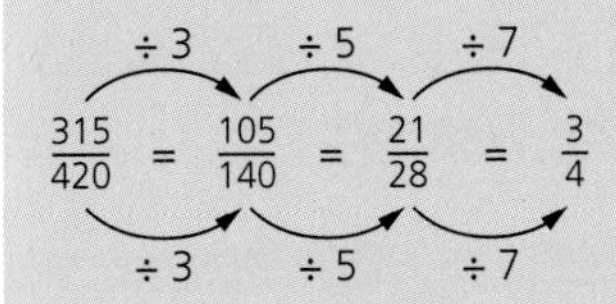

Pencil and paper multiplication and division

Believe me, if you know your tables up to 10 × 10, then you can multiply any two numbers up to 100 × 100. Actually you can multiply any two numbers as big as you like! Learning your tables thoroughly is probably the single most important skill in Mathematics.

Short multiplication usually refers to multiplying by a single-digit number. Long multiplication is used when multiplying by a number with two or more digits. Short division and long division are described in a similar way.

There are many methods of multiplying two numbers, and you probably know at least two. Look in the following section for some unusual methods that are fun to try. Why not research other methods from around the world?

Division always divides people into two groups. Those who get the right answer and those who do not. If you want to be in the first group, here are some things to remember:

- Practise by doing a multiplication question 'in reverse'. For example, if you have just worked out 12.3 × 4.8 = 59.04 try working out 59.04 ÷ 4.8 to see if you can get the right answer 12.3
- You can always get there by repeated subtraction if you are desperate (this method is sometimes known as 'chunking').
- Always check your answer by multiplying back again.
- Check that your answer is close to your original estimate.
- You can write any division question as a fraction and just cancel it down until you get your answer.

Decimals

Those who treat decimals as whole numbers are missing the point. Provided you put that point back again afterwards, though, it is the best way to do them.

How many decimal places do you make in the final answer?

When you *multiply* two decimals, you *add* the number of decimal places.

When you *divide* one decimal by another, you *subtract* the number of decimal places.

Example

6.71 (2 d.p.) and 1.1 (1 d.p.).

Multiplying one by the other:

$671 \times 11 = 7381$

so put in three decimal places (2 + 1) to get

$6.71 \times 1.1 = 7.381$

Dividing one by the other:

$671 \div 11 = 61$

so put in one decimal place (2 − 1) to get

$6.71 \div 1.1 = 6.1$

Remember once again to check by estimating: $7 \times 1 = 7$ and $7 \div 1 = 7$ so the answers we get by calculation should be similar, which they are.

Long multiplication (alternative methods)

It is useful to know a variety of methods for multiplying two numbers using pencil and paper.

Here are some interesting ways of finding 387 × 52 that are fun to try.

The Elizabethan method ('Napier's bones')

Write the two numbers across the top and down the right.

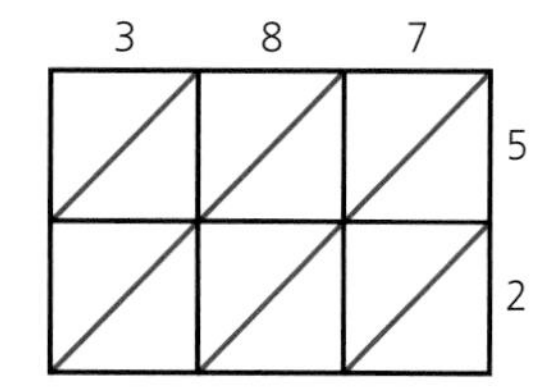

Multiply each pair of digits in the grid.

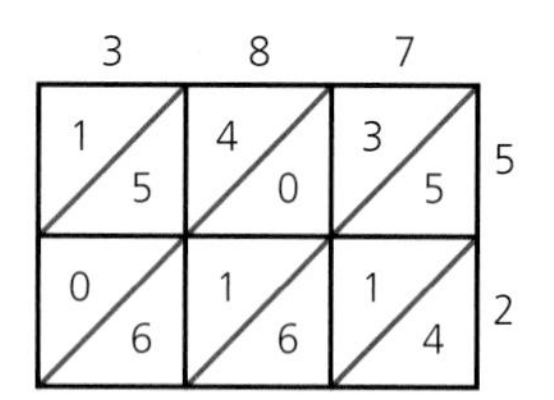

Add along the diagonals and the answer appears around the other corner!

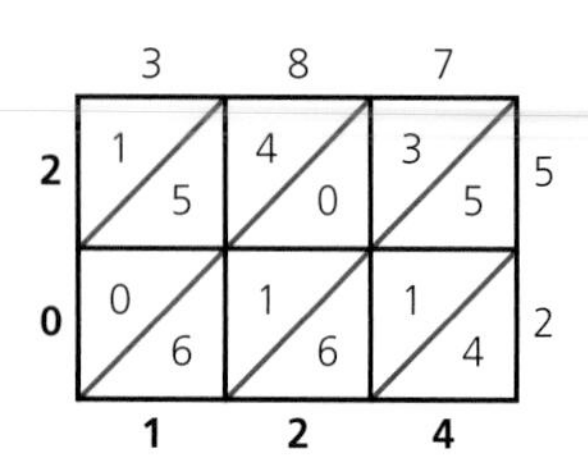

Add from the right and work backwards. Remember to carry the digits!

So 387 × 52 = 20 124

The Egyptian method

Make two columns. On the left write 1 and on the right write the larger number of the two. Keep doubling both. You get your answer by adding a clever selection of the rows.

1	387
2	774
4	**1548***
8	3096
16	**6192***
32	**12 384***

You only need to go this far because doubling 32 would give 64 and you do not need to go past 52 – the smaller number of the two.

Now choose the rows for which the *left* numbers add up to 52, then add up the numbers on the *right*.

*52 = 32 + 16 + 4

so just add up 12 384 + 6192 + 1548 = 20 124 which matches the previous answer.

The Russian method

Make two columns, with one number in each column.

Double the numbers on the right and halve the numbers on the left (ignoring remainders).

At the end, cross out the rows in which there is an *even* number on the left, then add up the remaining numbers on the *right*.

~~52~~	~~387~~
~~26~~	~~774~~
13	1548
~~6~~	~~3096~~
3	6192
1	12 384

Now there is a surprise! The same numbers are left as with the Egyptian method! Can you see why?

Using one calculation to find the answer to another

Did you know that you can use multiplication and division facts to calculate related problems?

Variations on the original problem are achieved using a combination of two procedures:

- Changing the order of the numbers in the question and answer.
- Multiplying or dividing any of the numbers involved by powers of 10 (i.e. 10 or 100 or 1000, etc.)

Example 1

Given that 5 × 7 = 35, *write down* the answers to these:

(The phrase *write down* means that you shouldn't need to do any detailed calculations.)

(i) 35 ÷ 7

(ii) 50 × 700

(iii) 350 ÷ 0.05

Answer

(i) 35 ÷ 7 = 5

This is just a restatement of the original problem, in a different order.

(ii) 50 × 700 = 35 000

If the number in the question is multiplied by 10, so is the answer.

This has been done three times here 5 × 10 and 7 × 10 × 10

so the answer is 35 × 10 × 10 × 10

(iii) 350 ÷ 0.05 = 7000

First use the fact that 35 ÷ 5 = 7 and think of the division as a fraction.

Then 350 ÷ 5 = 70

Multiplying the *numerator* by 10 multiplies the answer by 10

350 ÷ 0.5 = 700

Dividing the *denominator* by 10 multiplies the answer by 10

350 ÷ 0.05 = 7000

Dividing the denominator again by 10 multiplies the answer by 10 once more.

Alternatively, treat the division as a fraction to be simplified:

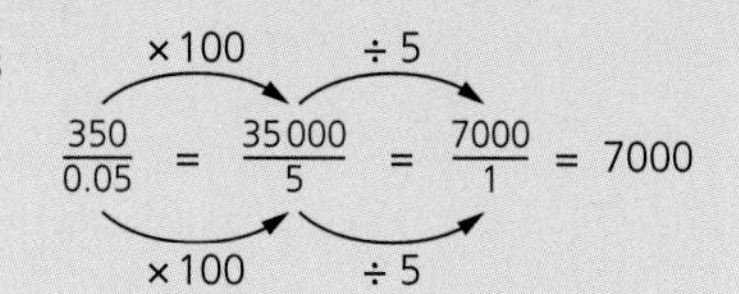

Example 2

Given that 58.7 × 89 = 5224.3, *write down* the answers to these:

(i) 5.87 × 8900

(ii) 52 243 ÷ 58.7

(iii) 5.2243 ÷ 0.089

Answer

(i) 5.87 × 8900 = 52 243

One number has been multiplied by 100 and the other has been divided by 10; the net result of this tug of war is × 10

(ii) 52 243 ÷ 58.7 = 890

We know that 5224.3 ÷ 58.7 = 89 (the original problem). Therefore 52 243 ÷ 58.7 is 10 times bigger.

(iii) 5.2243 ÷ 0.089 = 58.7

We know that 5224.3 ÷ 89 = 58.7 (the original problem). Both numbers in the division have been divided by 1000. If both numbers in a division are multiplied or divided by the same amount, then the answer stays the same. If you think about it, this is what equivalent fractions are all about!

× 1000

$$\frac{5.2243}{0.089} = \frac{5224.3}{89} = 58.7$$

× 1000

Exam-style questions

Try these questions for yourself. The answers are given near the back of the book.

Some of the questions involve ideas met in earlier work that may not be covered by the notes in this chapter.

Note: In this section you should not use a calculator. It is particularly important that you show all your working.

4.1 (a) Add 190.5 + 2909.6 (2)

(b) Subtract 2909.6 − 190.5 (2)

(c) Multiply 365 × 72 (2)

(d) Divide 24 144 ÷ 48 (2)

4.2 Evaluate, showing all your working:

(a) 3.95 + 0.471 (2)

(b) 3.95 − 0.471 (2)

(c) 7.12 × 0.5 (2)

(d) 7.12 ÷ 0.5 (2)

4.3 Evaluate:

(a) 16 − 9 ÷ 3 (1)

(b) 3 + 7 × 4 (1)

(c) 7 + 5 × 7 − 19 (2)

(d) 2 + 3 × (4 − 5) (2)

4.4 Given that 34 × 75 = 2550, find:

(a) 17 × 75 (1)

(b) 68 × 0.75 (2)

(c) 2550 ÷ 17 (2)

(d) 510 ÷ 3.4 (2)

4.5 By first writing each number correct to one significant figure, estimate the answer to:

(a) $\frac{3.19 \times 19.4}{9.8}$ (3)

(b) $\frac{7.98}{2.03 \times 3.92}$ (3)

4.6 (a) Calculate the total cost of buying 12 football shirts at £37.95 each. (3)

(b) When 16 people go to the theatre, the total cost of the tickets is £287.20

What is the cost of each ticket if they are all the same price? (3)

4.2 Calculator methods

Interpreting calculator displays

Money, time, distances, etc. – we know the difference but the poor calculator does not have a clue. It operates a simple RIRO rule (Rubbish In Rubbish Out), meaning that it cannot think for itself!

You could ask your calculator to work out £3 + 12 minutes − 4 bananas and it would tell you that the answer was 11, but that is just making the calculator look stupid.

Calculators sometimes give rounding errors, and you need to be careful with money and time answers. Always think of the context of the question.

3.7	£3.70 not 3 pounds and 7 pence.
4.25	4.25 hours = 4 hours 15 minutes (not 4 hours 25 minutes).
5.3	If we have worked out how many tins of paint we need then the answer is 6 not 5, since 5 would not be enough!
41.99999999999	This recurring decimal seems to suggest that the real answer is actually 42 after all!

Calculating fractions and percentages with a calculator

Did you know that every calculator has an OF button? No, not OFF – OF.

It looks like this: 

Let us see how it works.

Using a calculator to work with fractions

Some older fraction calculators have a button that looks something like this:

Example

Work out $\frac{2}{3}$ of 72

2 $a\,b/c$ 3	×	72	=	48
two-thirds	of	72	equals	48

If there is no fraction key on the calculator, we can use the division key instead:

2 ÷ 3	×	72	=	48
two-thirds	of	72	equals	48

Most modern calculators are equipped to deal with fractions more or less as they are written, and so they appear on screen almost exactly as they would appear on paper. When choosing a calculator for school and exam use, look for 'natural display' in its features. The most popular school calculators include this feature.

> Level 3
> For Level 3 candidates it is strongly recommended that you purchase a 'scientific' calculator. Do spend time reading the manual and trying the examples, however, and never buy a calculator just before an exam!

Remember

Percentages are just as easy. Remember that % means out of 100 and you have a fraction again, or a decimal if you prefer.

Example

Find 75% of 108

75 ÷ 100	×	108	=	81
75%	of	108	=	81

or simply

0.75	×	108	=	81
75%	of	108	=	81

Exam-style questions

Try these questions for yourself. The answers are given near the back of the book.

Some of the questions involve ideas met in earlier work that may not be covered by the notes in this chapter.

Notes:

- In this section a calculator is essential for most questions; you will be told if you should *not* use a calculator.
- Unless instructed otherwise, or unless the answer is exact, write all the figures shown in the calculator display first and then write the answer to three significant figures.

Remember:

- Different calculators work in different ways and you should have absolute confidence in your own calculator and your ability to use it.
- The calculator will respond faultlessly to the instructions given to it, so it is very important that you give it the appropriate instructions.
- In most cases, it is important to write more than just the 'answer' and you should say what you are doing.
- It is a good idea to know roughly what the calculator answer is likely to be before you start and to do a check after the calculation.

4.7 Calculate:

(a) $3.95 - {}^{-}7.83$ (1)

(b) $\dfrac{6.9 + 8.2}{3.9}$ (1)

(c) $\dfrac{6.45}{0.95 - 0.08}$ (1)

(d) $\dfrac{4.09 + 6.23}{3.97 \times 0.21}$ (1)

4.8 Calculate:

(a) $(29 + 413)^2$ (1)

(b) $\sqrt{11.45}$ (1)

(c) $\sqrt[3]{1.65}$ (1)

(d) $7\frac{1}{2}$ divided by $2\frac{3}{4}$, leaving your answer as a mixed number. (1)

4.9 Calculate:

(a) the area of a circle that has radius 5.8 cm (1)

(b) $47\,000 \times 80\,000$ (1)

4.10 Write an appropriate answer for each of these calculator displays:

(a) calculation involving money (£, GBP) (1)

10.8

(b) calculation involving time (in minutes) (1)

15.2

(c) calculation involving time (in hours) (1)

4.75

(d) calculation involving number of minibuses needed for a school trip (1)

3.8

(e) calculation involving fractions (1)

2 ⌋ 7

4.11 Write an appropriate answer for each of these displays from an older-style calculator:

(a) calculation involving fractions (1)

1 ⌋ 3 ⌋ 5

(b) calculation involving very large distances in miles (1)

4.5 ₉

(c) calculation involving small lengths in centimetres (1)

5.07 -03

(d) calculation involving sharing eight cakes between six people (1)

1.333333333

(e) calculation involving division (1)

9.999999999

4.12 (a) Writing down the full decimal display from your calculator, find the value of $\frac{27.3 - 5.4}{3.8}$ (2)

(b) Write your answer to part (a) correct to three significant figures. (1)

(c) Write your answer to part (a) correct to one decimal place. (1)

4.13 Brian's calculator display is shown here, after he has done a multiplication.

61.755

(a) Write the number shown in the display correct to two decimal places. (1)

(b) Write the number shown in the display correct to one significant figure. (2)

(c) Which number did Brian multiply by 3.45? (2)

4.14 (a) (i) Writing down the full decimal display from your calculator, evaluate (2)

$$\frac{123.9 \times 0.03}{207.5}$$

(ii) Write your answer to part (a) (i) correct to one significant figure. (1)

(b) (i) Writing down the full decimal display from your calculator, evaluate (2)

$$\frac{30.98}{371.5 - 456.5}$$

(ii) Write your answer to part (b) (i) correct to three decimal places. (1)

4.15 Look at this calculation:

$$\frac{287.3}{31.1 + 9.6}$$

(a) Without using a calculator, and showing all your working:

(i) rewrite the calculation shown above, giving each number correct to one significant figure (2)

(ii) evaluate your answer to part (a) (i). (1)

(b) (i) Using a calculator, and writing down the full decimal display, evaluate the original calculation. (1)

(ii) Write your answer to part (b) (i) correct to one significant figure. (1)

(iii) Write your answer to part (b) (i) correct to two decimal places. (1)

4.3 Checking results

Inverses and approximations as checks

It is a good habit always to check your answers by using inverses and approximations.

Inverses can be thought of as opposites. There are lots of opposite pairs in mathematics, but let us start with the most important four:

The inverse of this:	+	−	×	÷
is this:	−	+	÷	×

Checking using inverses

- We have just done the calculation 257 − 78 = 179

  We check the subtraction by *adding:* 179 + 78 = 257 OK!

- We have just done the calculation 108 ÷ 3 = 36

  We check the division by *multiplying:* 36 × 3 = 108 OK!

Estimation and approximation

Imagine that on your calculator you are working out 28.7 × 5.42

From the calculator display you obtain the answer 697.554

A quick approximation will tell you that this cannot be right!

28.7 is roughly 30, and 5.42 is roughly 5, so the answer must be about 150

Better try again!

Getting the correct answer

Checking your results often involves making use of mental strategies (see Section 3.2), including:

- making approximations to get an idea of the 'size' of the result
- using inverse operations (working backwards)
- looking at the units digit following a multiplication.

If your check suggests that an error has been made, this could be because you have:

- entered the wrong number into the calculator (very easy to do, particularly when you are not looking at the screen while you type)
- given the calculator inappropriate or incomplete instructions, such as failing to press = at the right time
- set out written calculations incorrectly, such as forgetting 'carrying digits' or failing to take account of place value
- made 'zero' errors, such as leaving them out or putting in extra zeros
- made a simple 'times tables' or other computation error.

Of course it is best to avoid making errors in the first place.

You are less likely to make errors if you:

- have a rough idea of the result before you start
- set out written working clearly
- write down intermediate calculator results.

A final word of caution!

If your check suggests that you may have made an error, do not forget that the error may be in your check!

If your check confirms that your original answer is likely to be correct, then that is probably fine. If your check casts doubt upon your original answer then it is safer to check again before changing your answer!

Exam-style questions

Try these questions for yourself. The answers are given near the back of the book.

Some of the questions involve ideas met in earlier work that may not be covered by the notes in this chapter.

Note: Questions 4.16 and 4.17 involve the use of mental strategies and should be done entirely in your head – write down only the answer.

4.16 (a) I invest £800 at 5% interest. How much interest do I earn in a year? (1)

(b) Tommy played in four soccer matches. The figures show the number of goals he scored in each match. How many goals does he need to score in his next match to make his mean number of goals four per match? (1)

3 0 5 2

(c) A recipe says that 6 kg of flour will make 12 loaves of a certain size. How many kilograms of flour do I need to make 20 loaves of the same size?(1)

(d) On a trip to Edinburgh, Len spends £14.40 on the train, £8.00 on a taxi and £2.60 on a bus. How much does the journey cost him altogether? (1)

(e) The information shows the charges for parking in a car park for different lengths of time.

0–2 hours 90p

each extra hour
or part hour 40p

If I arrive at 09:30 in the morning and leave at 13:45 the same day, how much do I pay to park? (1)

4.17 **(a)** This calculation is used to find the total cost in pounds of some items in a sale.

$$\frac{11 \times 15}{0.3}$$

Calculate the total cost. (1)

(b) Tina is 13 years old. Five years ago her sister was half Tina's age. How old is Tina's sister now? (1)

(c) An A6 sheet of paper, a quarter the size of an A4 sheet, has an area of about 150 cm². What approximate area in square metres could be covered by 100 sheets of A6 paper? (1)

(d) How many cube-shaped boxes with edges of half a metre can I fit into a case measuring 4 m by 2 m by 1 m? (1)

(e) The circumference of Ed's cycle wheel is 249 cm. Estimate, to one significant figure, how many times the wheel goes round while he cycles 10 km. (1)

4.18 When Mrs Watt goes shopping she keeps a mental note of the approximate total of the items as she puts them into the trolley. If the item cost is less than 49p she ignores it, if it costs 50p to £1.49 she counts it as £1, if it costs £1.50 to £2.49 she counts it as £2 and so on.

Here is one of Mrs Watt's shopping bills.

SHERRY	£5.39
MILK	£0.86
6 EGGS	£1.50
TIGER ROLLS	£0.55
MILK	£0.45
FELIX POUCH	£3.43
FELIX POUCH	£3.43
REDUCED ITEM	£1.01
REDUCED ITEM	£0.39
PUN PEACH	£1.00
SHORTBREAD	£0.81
BISCUITS	£2.93
REDUCED ITEM	£0.91
GO CAT	£2.55
GOCAT COMP V	£2.36
BABY ACC	£0.70
HANDWASH	£1.00

(a) Use Mrs Watt's mental strategy to find the approximate total cost of the items in her trolley. (2)

(b) Use your calculator to find the exact cost of the items. (2)

4.19 (a) Julian has thought of a number. When he adds 3 to his number, then multiplies by 4 and finally subtracts 6, he gets 30

Megan says that the number Julian thought of is 6

By means of a flow chart showing the inverse operations, or otherwise, show Megan how to check. (3)

(b) Moira has thought of a number. When she subtracts 4 from her number, then divides by 3 and finally adds 11, she gets 45

John says that the number Moira thought of is 109

Check to see if John is correct and, if he is not, say what Moira's number is. (3)

★ Make sure you know

- ★ All of the material in Chapter 3
- ★ How to use all four operations with decimals to two places (including rounding to 1 or 2 decimal places
- ★ An appropriate non-calculator method for solving problems that involve multiplying and dividing any three-digit number by any two-digit number
- ★ The effects of multiplying and dividing by numbers between 0 and 1
- ★ How to solve numerical problems involving multiplication and division with numbers of any size, using a calculator efficiently and appropriately
- ★ How to check your solutions by applying inverse operations or estimating, using approximation
- ★ How to use the glossary at the back of the book for definitions of key words

Test yourself

Before moving on to the next chapter, make sure you can answer the following questions.

The answers are near the back of the book.

1 Showing all your working, evaluate:

(a) $6.04 + 3.5$

(b) $6.04 - 3.5$

(c) 6.04×3.5

(d) $6.04 \div 4$

2 Showing all your working, evaluate:

(a) $31.47 + 17.58$

(b) $31.47 - 17.58$

(c) 6.79×0.35

(d) $6.79 \div 0.35$

3 Showing all your working, calculate:

(a) $4 + 5 \times 6$

(b) $(10 \times 4.99) + (5 \times 0.85) + (3 \times 2.05)$

(c) $3 \times 4 + 6 - 5 \times 2$

4 (a) Showing all your working, calculate

(i) $3.6 + 6.6 \div 5$

(ii) $1.9 + 2.9 \times (3.9 - 4.9)$

(b) What is the difference between your two answers in part (a)?

5 (a) Look at this calculation: $\dfrac{39.7}{8.9 \times 9.1}$

Without using a calculator, and showing all your working:

(i) rewrite the calculation shown above, giving each number correct to 1 significant figure

(ii) evaluate your answer to part (a) (i).

(b) (i) Now, using a calculator and writing down the full decimal display, evaluate this calculation.

(ii) Write your answer to part (b) (i) correct to two significant figures.

(iii) Write your answer to part (b) (i) correct to three decimal places.

6 (a) (i) Writing down the full decimal display from your calculator, find the value of

$$\frac{47.9}{8.1 \times 0.8}$$

(ii) Write your answer to part (a) (i) correct to three significant figures.

(b) (i) Writing down the full decimal display from your calculator, find the value of $20.7 \div (9.37 - 2.6)$

(ii) Write your answer to part (b) (i) correct to three decimal places.

Note: Questions 7 and 8 involve the use of mental strategies and should be done entirely in your head – write down only the answer.

7 (a) In Rugby Union, a converted try is worth seven points. A penalty is worth three points. In a match, a team scores two converted tries and four penalties. What is the team's total score?

(b) The total cost of a coach trip for 20 people is £2300. How much will each person pay if the cost is divided equally between them?

(c) Tiles are sold in packs of ten and cost £9 per pack. I estimate that a bathroom requires 150 tiles. How much will the tiles cost?

(d) A pudding that feeds four people requires 240 ml of milk. How much milk is required if the pudding is to feed six people?

(e) Find the value of 99^2

8 (a) A fire destroyed 60 out of the 240 paintings in a gallery. What percentage of the original number of paintings was destroyed by the fire?

(b) The masses of three packets are 55 g, 35 g and 60 g. What is the mean mass of the packets?

(c) The cost to enter an adventure park is £16 per person. There is a reduction of one quarter for parties of ten or more. What is the total cost for a group of ten people to go to the adventure park?

(d) Calculate, in hectares, the area of the rectangular field shown.

300 m

150 m

not to scale

1 hectare = 10 000 square metres

(e) I need to attend a meeting at my bank, which is 10 minutes' walk from the bus stop in Dimwich, at half past ten. What is the latest time I can catch the bus in Waledale?

Waledale depart	0740	0820	0910	1000
Ploptown arrive	0825	0905	0955	1045
Dimwich arrive	0855	0935	1025	1115

5 Problem solving (1)

Note: The material in this chapter is not detailed in the ISEB syllabus but it is included here to help with learning and practising important skills needed in examinations.

5.1 Decision making

Choosing a strategy

When tackling a problem, it is important to decide on a suitable strategy. This is largely a matter of common sense, but you could consider the following strategies.

Making an organised list (listing all possible outcomes)

Example

How many different integers (whole numbers) could you make using the three number cards here?

There are obviously three single-digit numbers: 1, 2 and 3

There are six two-digit numbers: 12, 13, 21, 23, 31, 32

There are also six three digit numbers: 123, 132, 213, 231, 312, 321

So, altogether, you could make 15 different integers.

Trying a practical approach (drawing diagrams, etc.)

Example

How many diagonals does a regular pentagon have?

Doing a rough sketch should help you.

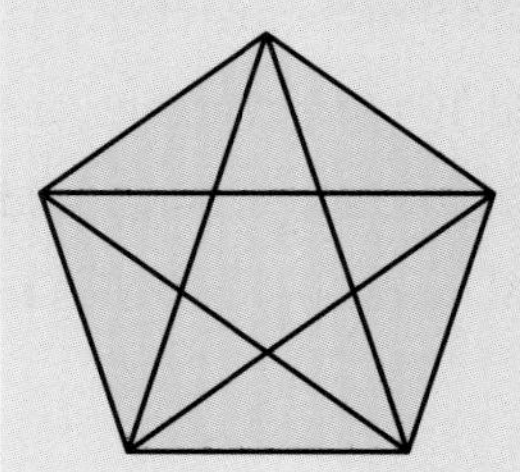

Guessing and checking (trial and improvement)

Example

Amy is thinking of three different positive integers and has given these clues:

'The product of my numbers is 72 and their sum is 18'

What are Amy's numbers?

You know that 72 is 2×36 and 36 is 3×12, so you could guess that the numbers are 2, 3 and 12

Unfortunately, this cannot be correct since the sum of 2, 3 and 12 is 17, not 18

What would be your next guess?

Trying a simpler example (trying easier numbers etc.)

Example

How many pound coins could you place in a single layer on a tennis court?

You might like to start by seeing how many pound coins you could fit into a 10 cm square. It would also be a good idea to see how best to arrange the coins – rows and columns may not be the best way!

Looking for a pattern

Example

Here are some 'steps' made from matchsticks.

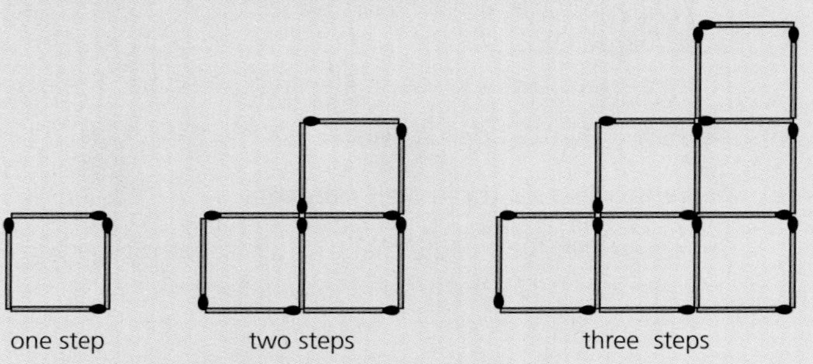

How many matches would be needed to make six steps?

One step has 4 matches.

See how many matches must be added to make two steps: we need 6 more matches, making 10 altogether.

To make three steps we need to add 8 more matches, making a total of 18

To make four steps we need to add 10 more matches, and so on.

You will see that there is a pattern here: add 6, add 8, add 10, add ...

Follow the pattern to find the total number of matches for six steps.

Exam-style questions

Try these questions for yourself. The answers are given near the back of the book. The questions in this chapter are not reflective of typical examination questions as the material is more concerned with practising important skills.

Some of the questions involve ideas met in earlier work that may not be covered by the notes in this chapter.

5.1 Decide which method you might best use to solve each of the following problems.

You do not need to solve the problems! Just write M, MJ, PP or C.

- **M**ental – entirely in your head
- **M**ental with **J**ottings – perhaps writing down an intermediate result
- **P**encil and **P**aper – setting out a calculation (such as a long multiplication) on paper, or doing a drawing
- **C**alculator – only if all other methods are unsuitable!

(a) What is the cost of ten stamps costing 61 pence each? (1)

(b) How many diagonals does a heptagon have? (1)

(c) What is the product of 47 and 29? (1)

(d) Which number multiplied by itself is 841? (1)

(e) How many DVDs costing £8.99 each could you buy for £100, and how much change would you receive? (1)

5.2 Which equipment would you use, and how would you use it, to solve these practical problems?

(a) Find the height of a friend. (2)

(b) Find the distance around your waist. (2)

(c) Find the temperature of water in a washing up bowl. (2)

(d) Find the volume of water in a washing up bowl. (2)

(e) Find the length of time it would take you to hop 100 metres. (2)

(f) Find the mass of a baked bean. (2)

(g) Find the area of the sole of your foot. (2)

5.2 Reasoning about numbers or shapes

When solving puzzles and problems about numbers or shapes, you may need to:

- make use of a variety of strategies
- make use of your knowledge and experience
- make careful observations
- do more thinking than writing
- ask yourself questions, such as 'What if ...'
- be prepared to try a different strategy if you seem to be getting nowhere.

Exam-style questions

Try these questions for yourself. The answers are given near the back of the book.

Some of the questions involve ideas met in earlier work that may not be covered by the notes in this chapter.

5.3 The sum of two numbers is 109

One number is 9 less than the other.

What are the two numbers? (2)

5.4 Gina has three coins in her pocket. She has a total of 62 pence.

(a) What are the three coins? (1)

She buys a pen costing 53 pence, handing over two of her coins.
The shopkeeper gives her two coins as change.

(b) Which coins does the shopkeeper give her? (2)

5.5 Robbie has thought of a number and has given these clues.

The number

- is smaller than 50
- is one less than a multiple of 6
- is *not* prime.

What is Robbie's number? (3)

5.6 Lynn has thought of two numbers.

The product of her numbers is 36

The difference between her numbers is 5

What is the sum of her numbers? (3)

5.7 **(a)** The shapes A, B and C have been made by joining together two congruent tiles like this one.

A B C

(i) Which shape has reflection symmetry only? (1)

(ii) Which shape has rotation symmetry only? (1)

(iii) Which shape has no symmetry? (1)

(b) On a square dotted grid or squared paper, using the same simple tile, draw your own shapes, D with reflection symmetry and E with rotation symmetry. You may use more than two tiles if you wish. (4)

5.3 Real-life mathematics

Everyday situations

Many examination questions concern everyday situations including, for example:

- shopping
- DIY tasks
- holidays
- journeys
- cooking with recipes.

The mathematical knowledge and skills involved can be from all areas including, for example:

- money and currency conversions
- fractions, decimals and percentages
- measurement
- ratio
- bearings
- data tables, graphs, etc.

Exam-style questions

Try these questions for yourself. The answers are given near the back of the book.

Some of the questions involve ideas met in earlier work that may not be covered by the notes in this chapter.

5.8 Janet has £4.20 and John has £1.40 more than Janet. They decide to buy a DVD costing £6.80 and share the money that is left equally.

How much money will each of them have when they have paid for the DVD? (3)

5.9 Mrs Smith buys three pints of milk costing 43 pence per pint, two loaves of bread costing 78 pence each and 500 grams of cheese costing £1.10

How much change will she receive from a £10 note? (3)

5.10 The information on a caramel bar is

mass per 100 g	
protein	6 g
fat	20 g
carbohydrate	54 g

(a) What fraction of the bar is fat? (2)

(b) What percentage of the bar is carbohydrate? (1)

A caramel bar has a mass of 55 grams.

(c) What will be the mass of fat in a bar? (2)

5.11 The 16 pupils in a class make biscuits in a cookery lesson.

Here is the recipe that each pupil uses to make 12 biscuits:

plain flour	180 g
cornflour	45 g
butter	215 g
icing sugar	80 g

(a) How many biscuits will the class make altogether? (2)

(b) (i) How much icing sugar will be needed altogether? (2)

(ii) The teacher opens a 2 kg packet of icing sugar. How many grams of icing sugar will be left at the end? (2)

(c) What is the total mass of ingredients used by each pupil? (2)

(d) The biscuits lose 15% of their mass during the cooking process. What will be the mass of one pupil's cooked biscuits? (2)

(e) The cost of the ingredients for each pupil is 72 pence. What is the cost of each biscuit? (1)

(f) Half of the children sell their biscuits for 5 pence each at a school charity event. How much money will they raise altogether? (2)

★ Make sure you know

- ★ How to develop your own strategies for solving problems and use these strategies
- ★ How to look for patterns and relationships
- ★ How to search for a solution by trying out ideas of your own
- ★ How to present information and results in a clear and organised way
- ★ How to check that your results are reasonable
- ★ How to use your knowledge, understanding and skills, often in several areas of mathematics, to tackle everyday problems
- ★ How to use the glossary at the back of the book for definitions of key words

Test yourself ✓

Before moving on to the next chapter, make sure you can answer the following questions.

The answers are near the back of the book.

1 Find the area of your hand.

2 Rhona has thought of a number and has given these clues.

The number:

- is larger than 10 but smaller than 60
- has a units digit that is one less than the tens digit
- is prime.

What is Rhona's number?

3 Jo has the three tiles shown below.

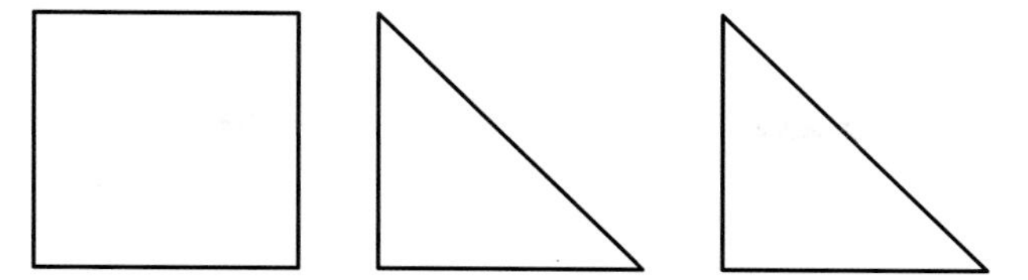

She arranges the tiles edge to edge to make the following shape.

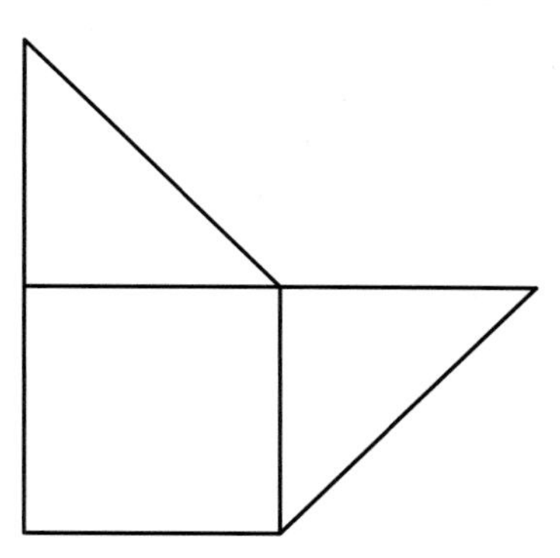

See how many different shapes you can make using all three tiles.

Note that non-exact edge fits and congruent shapes as shown below don't count.

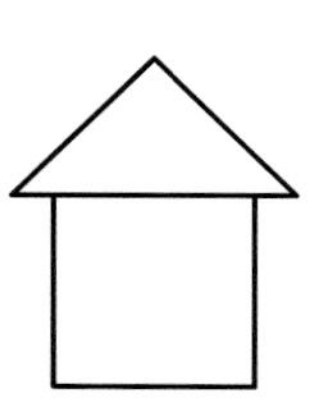

non-exact edge fit

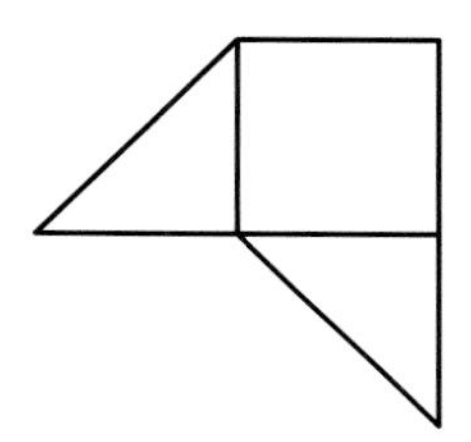

congruent shapes

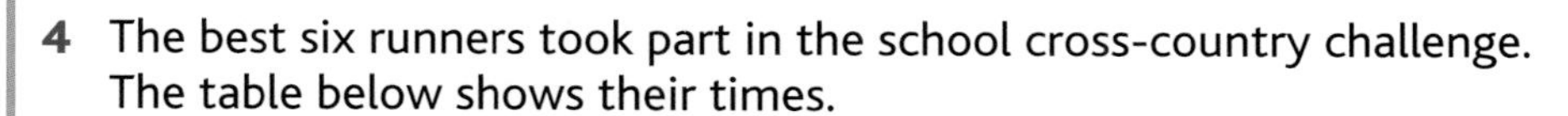

4 The best six runners took part in the school cross-country challenge. The table below shows their times.

Angus	51 minutes 2 seconds
Brenda	48 minutes 43 seconds
Clare	50 minutes 27 seconds
David	48 minutes 58 seconds
Ed	49 minutes 4 seconds
Flora	49 minutes 11 seconds

(a) (i) Who won?

(ii) Who finished last?

(b) What was the difference between the first and last runners' times?

Unfortunately Flora fell in the stream and she lost 20 seconds scrambling out.

(c) Had Flora not fallen in the stream, in what position would she have finished?

5 Janet had a good idea for raising money for charity. She bought 200 plastic balls of different colours. The balls were identical apart from the colour.

The table shows the numbers of the balls of each colour.

The number of orange balls was the same as the number of purple balls.

The number of green balls was three times the number of orange balls.

Colour	Number of balls
Red	40
Blue	25
Yellow	35
Green	
Orange	
Purple	

(a) How many green balls were there?

Janet paid £20 for the balls and sold all of them at 15p each.

(b) (i) How much was the cost of each ball?

(ii) How much profit did Janet make altogether?

The purchasers wrote their names on the balls and then the balls were dropped at the same time from a bridge over the narrow river. Everyone ran to the next bridge 100 metres downstream and Janet stood ready to fish the winning ball from the water.

(c) Which colour ball was most likely to win?

Gail bought one ball of each colour and Stephanie bought six green balls.

(d) Who was the most likely to have the winning ball, Gail or Stephanie?

6 Problem solving (2)

The material in this chapter is not detailed in the ISEB syllabus but it is included here to help with learning and practising important skills needed in examinations.

6.1 Decision making

Generalising and testing

Generalising and testing usually involves forming a 'rule' and then testing it before using it to solve the problem.

We will look again at the 'steps' problem from Section 5.1 in Chapter 5.

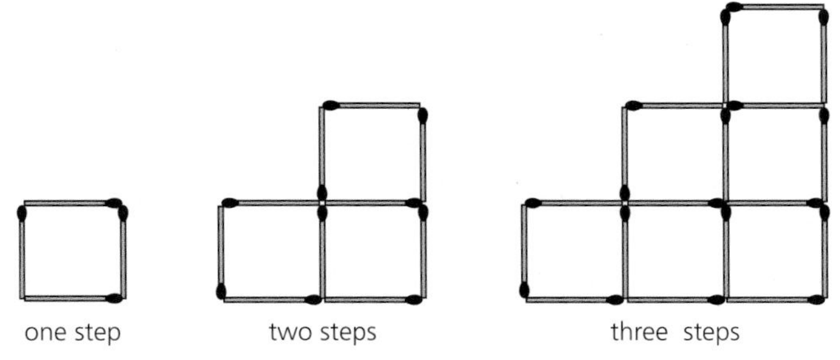

How many matches would be needed for 100 steps?

We could simply follow the pattern described in Chapter 5 but clearly this would take a long time!

We can draw up a table, where the number of steps is n, as follows:

n	1	2	3	4	5	6
n^2	1	4	9	16	25	36
$3n$	3	6	9	12	15	18
$n^2 + 3n$	4	10	18	28	40	54

The expression $n^2 + 3n$, gives us the numbers of matches in the different steps.

This gives us the formula (rule) $m = n^2 + 3n$, where n is the number of steps and m is the number of matches.

We can test the formula by applying it to seven steps. If it doesn't work, we need to think again!

We can then use the formula to find the solution to the problem. This is much quicker! When n is 100, $n^2 + 3n$ is 10 300

You will find more information about this in Chapter 8.

Exam-style questions

Try these questions for yourself. The answers are given at the back of the book. The questions in this chapter are not necessarily reflective of typical examination questions as the material in this chapter is more concerned with practising important skills.

Some of the questions involve ideas met in earlier work that may not be covered by the notes in this chapter.

6.1 Decide which strategy you would use to solve each of the following problems.

You do not need to solve the problems!

- Make an organised list.
- Try a practical approach (such as drawing a diagram).
- Guess and check.
- Try a simpler example first.
- Look for a pattern.
- Find a formula and then test it before using it.

(a) How many diagonals does a regular nonagon have? (1)

(b) What is the most likely score you could get by adding together the numbers when these two spinners are used together? (1)

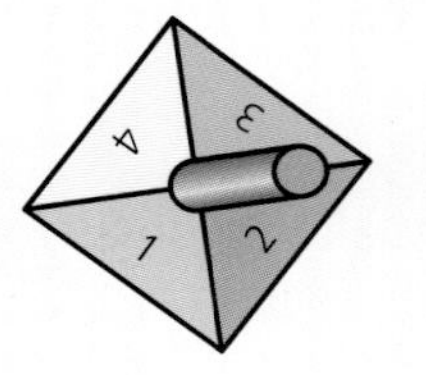

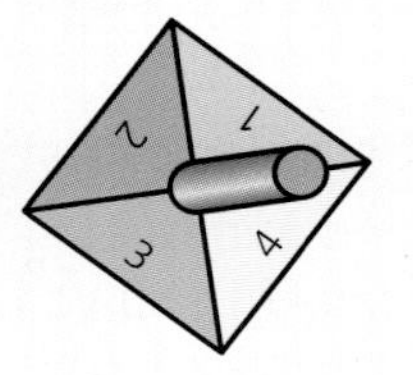

(c) Without using a calculator, find the square root, to one decimal place, of 30 (1)

(d) Find the next two terms of this sequence. (1)

1 6 16 36 ...

(e) How many different routes are there through this maze? (1)

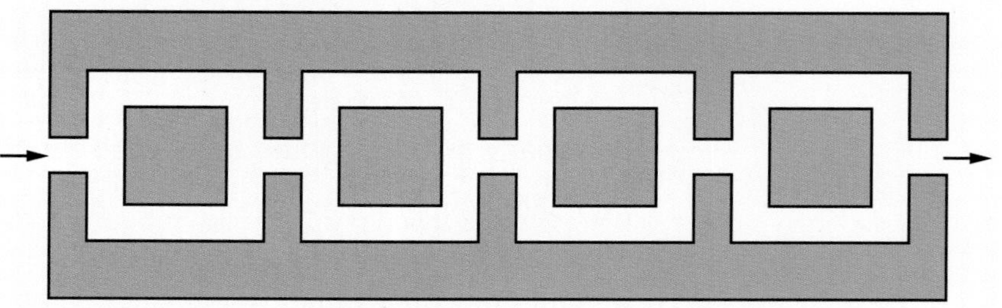

(f) Find the hundredth term of this sequence. (1)

3 8 13 18 ...

6.2 Reasoning about numbers or shapes

Triangular numbers

At the start of a game of snooker there are 15 reds in the triangle.

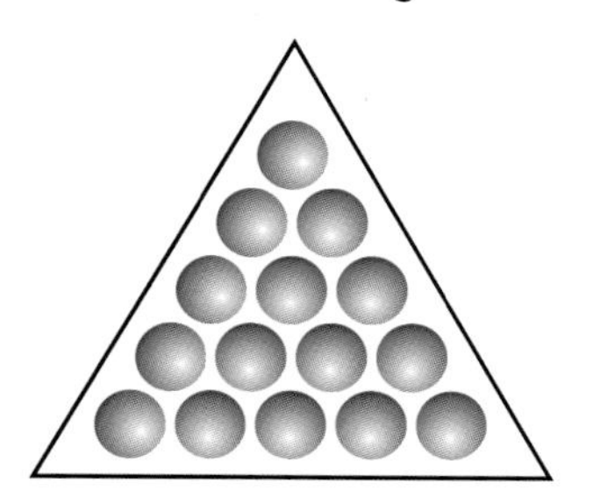

Which other numbers of balls could be arranged in a triangle this way?

The answer is a sequence of numbers that we call the triangular numbers.

They come up a lot in mathematics. For example, the number of handshakes required when everyone in a group of n people shakes hands with everyone else – it is always a triangular number!

Here are the first ten triangular numbers:

n	T_n
1	1
2	3
3	6
4	10
5	15
6	21
7	28
8	36
9	45
10	55

In the table, T_n simply means the nth triangular number.

Notice how they go up:

+ 2, + 3, + 4, + 5, etc. added on to the one before.

Also notice that any two consecutive *triangular* numbers give a sum that is a *square* number, e.g. 21 + 28 = 49

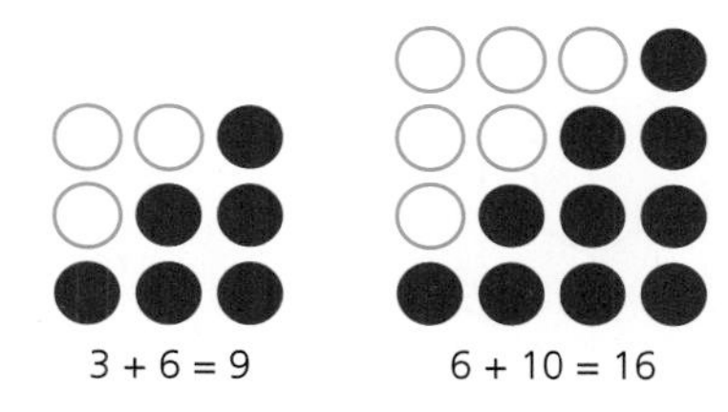

In general though, we can jump to any triangular number we want simply by using this formula:

$$T_n = \frac{1}{2}n(n+1)$$

Exam-style questions

Try these questions for yourself. The answers are given at the back of the book.

Some of the questions involve ideas met in earlier work that may not be covered by the notes in this chapter.

6.2 A plant produces three flowers in its first year. Every year after that, it produces two more flowers than in the previous year, as shown in this diagram.

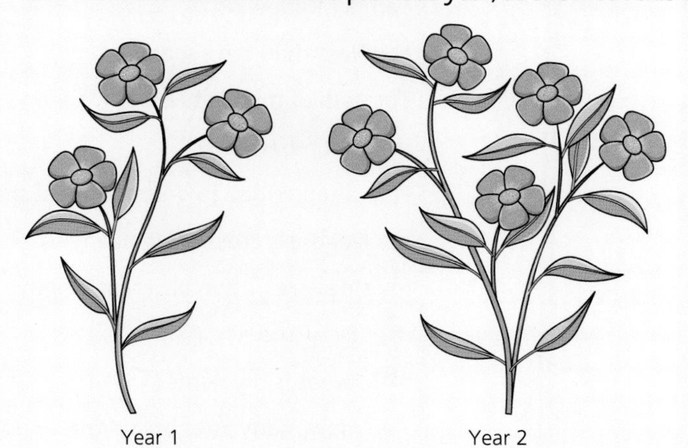

(a) Complete the table below to show the number of flowers in each year. (2)

Year	Number of flowers
1	3
2	5
3	7
4	
5	

(b) How many flowers will the plant have in its seventh year? (2)

(c) Assuming that the plant continues to grow at the same rate, during which year will the plant produce more than 30 flowers for the first time? (2)

6.3 The diagram shows a cuboid made up of small cubes.

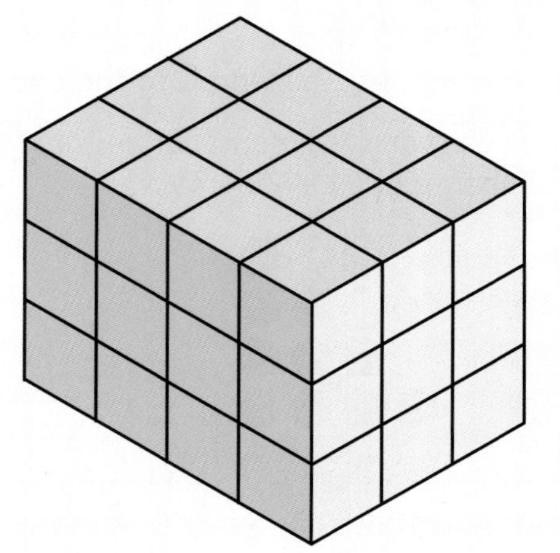

(a) How many small cubes have been used to make the cuboid? (2)

The whole outer surface of the cuboid is painted blue.

(b) How many of the small cubes have

- **(i)** only one face painted blue (2)
- **(ii)** only two faces painted blue (2)
- **(iii)** three faces painted blue (2)
- **(iv)** no faces painted blue? (1)

6.4 Answer the following puzzles:

(a) I am a multiple of 7

I am less than ten squared.

The sum of my digits is 10

What number am I? (2)

(b) I am a four-sided two-dimensional shape.

My diagonals cross at right-angles.

My sides are of two different lengths.

- **(i)** Draw me. (2)
- **(ii)** What is my name? (1)
- **(iii)** How many lines of symmetry do I have? (1)

6.5 Here are the first three patterns in a sequence, using 1 cm lines drawn on dotted paper.

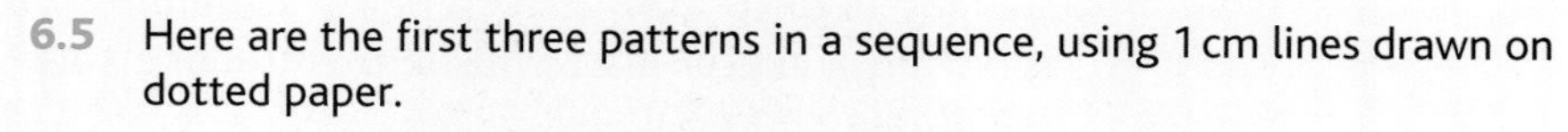

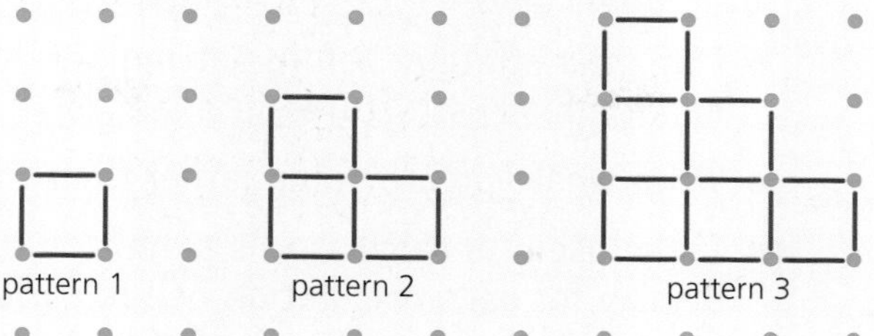

(a) Draw the next pattern in the sequence. (2)

(b) Extend the following table of values for pattern numbers 3 to 5 (6)

Pattern number	Number of squares	Total number of dots	Perimeter (cm)
1	1	4	4
2	3	8	8

(c) By considering the sequence of numbers in each column of the table from part (b), extend the table to include values for pattern number 8 (3)

6.3 Real-life mathematics

An examination question can cover a number of syllabus areas and require a wide range of strategies.

Many questions are concerned to some extent with money or with fractions, decimals and percentages.

The real-life situations may include:

- shopping – money, discounts
- cooking and baking – recipes
- eating and diets – calories
- medicines and health – measures
- growth – measurement, line graphs
- exercise and sports – times, speeds, scores
- holidays and travel – driving, exchange rates
- planning events – hiring equipment, fundraising
- gardening – weed killing, filling a pond
- DIY tasks – building a shed, decorating a room
- business and banking – VAT, profit and loss, interest rates
- design – tessellations, symmetry
- exploring – maps, scales
- model-making – scales, nets
- puzzles.

Use these general hints to help you solve a problem:

- Since questions can be rather 'wordy', read each question very carefully to make sure that you understand what is required. Remember that the little word 'not' changes the meaning considerably!

- It may help to draw a sketch. For example, if a scale drawing is required, drawing a sketch first will help you make sure your drawing fits into the available space.
- Plan your strategy before putting pen to paper, particularly if the question is not already structured into stages.
- Write what you are doing rather than simply jotting down calculations.
- If the use of a calculator is permitted, then write down the full calculator display for each calculation before correcting it to three significant figures for a final answer. When several calculations are needed, always use the full value in your calculator for later calculations.
- Pay attention to units throughout and always check that your final answer is in the correct units.
- When you have finished, ask yourself: 'Can the marker follow through this and see exactly what I have done?' In the event that your answer (for whatever reason) is incorrect, then the marker will still be able to award marks for the written evidence of your thinking.

Exam-style questions

Try these questions for yourself. The answers are given near the back of the book.

Some of the questions involve ideas met in earlier work that may not be covered by the notes in this chapter.

6.6 (a) Estimate the cost of 49.8 litres of petrol at 101.1 pence per litre. Give your answer to the nearest pound. (2)

(b) On a journey of 450 miles, my car uses 1 litre of petrol every 9.1 miles. Estimate the number of litres of petrol that my car uses on the journey. (2)

6.7 (a) Alice went to the shop and bought seven DVDs costing £5.99 each. How much did Alice spend on the DVDs? (2)

(b) (i) Max bought five books, three costing £6.99 each and the other two costing £3.49 each.

What was the total amount that Max spent? (2)

(ii) How much change should Max receive from a £50 note? (2)

(c) Mr Lester purchased 28 identically priced theatre tickets for a school trip, at a total cost of £474.60
What was the cost of each ticket? (2)

6.8 Flora intends to bake 18 small cakes for her party. She looks up a recipe in an old cookery book and finds that to make a dozen cakes she needs the following ingredients:

flour 8 ounces

margarine 6 ounces

sugar 4 ounces

(a) Write down the amount of each ingredient that Flora needs to bake the 18 cakes. (3)

Flora finds that she has only 10 ounces of flour, but plenty of the other ingredients.

(b) What is the largest number of cakes she can bake? (2)

Revision tip

If you are interested in becoming more skilled in problem solving, I can recommend a classic book by George Polya called *How To Solve It*. It contains some excellent general advice.

★ Make sure you know

★ All of the ideas in Chapter 5

★ How to show understanding of situations by describing them mathematically using symbols, words and diagrams

★ How to draw simple conclusions of your own and explain your reasoning

★ How to check your working and results, considering whether they are sensible

★ How to carry out substantial tasks and solve quite complex problems by breaking them down into smaller, more manageable tasks

★ How to give mathematical justifications, making connections between the current situation and situations you have met before

★ How to use your knowledge, understanding and skills, often in several areas of mathematics, to tackle everyday problems

★ How to use the glossary at the back of the book for definitions of key words

Test yourself

Before moving on to the next chapter, make sure you can answer the following questions. The answers are at the back of the book.

In question 6.1 you decided upon *strategies* for solving six very different problems.

In the first six questions here, you are asked to *solve* those problems.

1 How many diagonals does a regular nonagon have?

2 What is the most likely score you could get by adding together the numbers when these two spinners are used together?

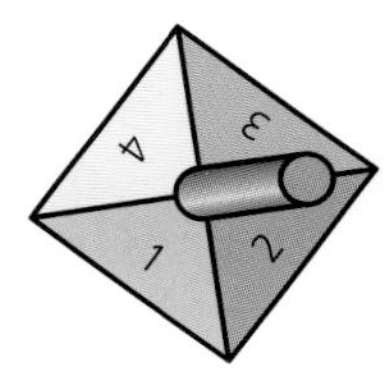

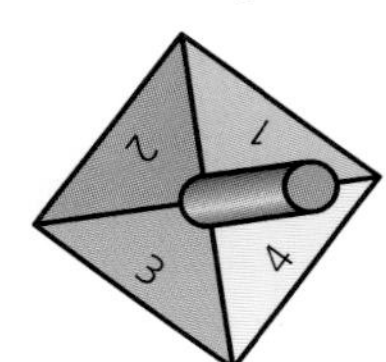

3 Without using a calculator, find the square root, to one decimal place, of 30

4 Find the next two terms of this sequence.

1 6 16 36 …

5 How many different routes are there through this maze?

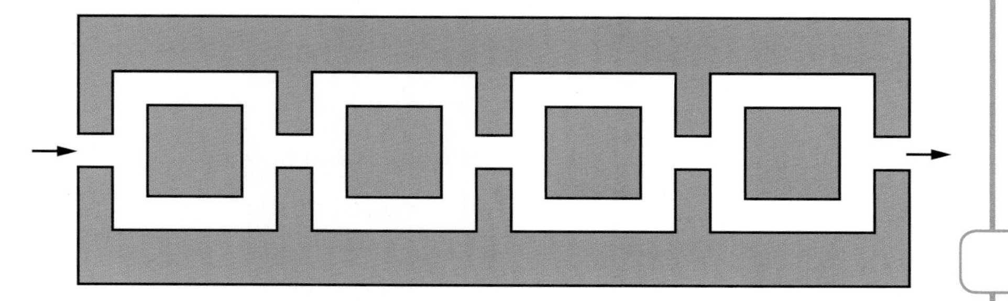

6 Find the hundredth term of this sequence.

3   8   13   18 ...

7 Aunt Jane has divided her bear collection between her three nieces as follows:

Gail has $\frac{1}{3}$ of the total.

Hilda has $\frac{1}{4}$ of the total.

Isla has the rest.

(a) What fraction of the total does Isla have?

Gail divides her share equally between her twin brothers, Jack and Kane.

(b) What fraction of the total does Jack have?

(c) If Kane now has four of Aunt Jane's bears, how many bears were there in Aunt Jane's collection?

8 George stacks shelves at the supermarket.

He earns £7 per hour for the first 32 hours he works in a week.

(a) How much is he paid for a 32-hour working week?

George is paid £8.50 per hour for every extra hour he works over 32 hours.

(b) One week he worked for a total of 40 hours.
How much did he earn in that week?

(c) Another week he earned £334.50
How many hours did he work that week?

9 A shallow tray is made from a rectangular sheet of card.

Small squares of side 2 cm are removed from the corners, as shown, to make the net.

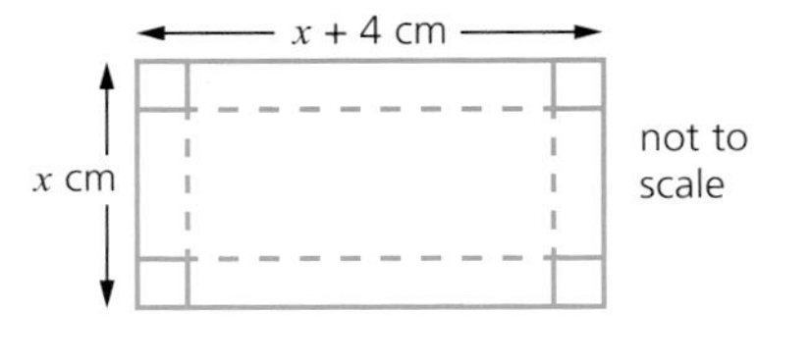

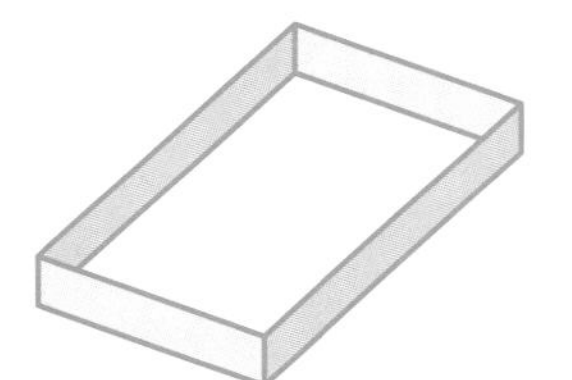

The card is folded along the dotted lines to make the tray.

(a) What is the height of the tray?

(b) Write down, in terms of x, expressions for:

(i) the length of the tray

(ii) the width of the tray

(iii) the surface area of the inside of the tray.

(c) If I start with a sheet of card measuring 24 cm by 20 cm, find

(i) the surface area of the inside of the tray

(ii) the volume of the tray.

10 A useful indicator of appropriate human body mass (and therefore weight) is the body mass index (BMI).

BMI can be found using the formula:

$$\text{BMI} = \frac{\text{mass}}{\text{height}^2}$$

where the mass is in kilograms and the height is in metres.

For example, if a man is 1.80 metres tall and weighs 72 kg, his BMI is $\frac{72}{1.8^2} = 22.22$ to two decimal places.

(a) Calculate the BMI of the following people. Give your answers to two decimal places:

(i) Angus is 1.65 m tall and weighs 67 kg.

(ii) Beatrice is 1.78 m tall and weighs 83 kg.

(iii) Colin is 1.59 m tall and weighs 66 kg.

The range of BMI values for adults covers various categories, including:

Category	Underweight	Normal	Overweight
BMI	<18.5		>25

If an adult's BMI is between 18.5 and 25, then he/she is considered to be in the normal range.

(b) Who is overweight in part (a)?

(c) If a man is 1.85 metres tall, calculate his maximum mass (to 2 d.p.) before he is considered to be overweight.

7 Algebra (1)

7.1 Equations and formulae

Word formulae and function machines

Simple **formulae** can be expressed in words or as a picture.

How many months old are you?

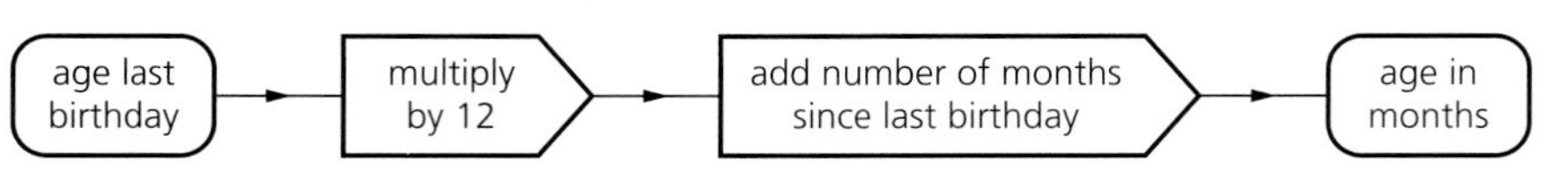

This is an example of a **word formula**, written as a simple **flow chart**, to change years and months into just months.

A word formula is a set of instructions telling you how to change one number into another.

Here's another type of word formula, written as a puzzle:

I think of a number, double it, add 11 and the answer is 99

What was my number?

Revision tip

Get a friend to make up some word puzzles like this for you. Discuss how you can solve each one by working backwards, and then check your answer by putting it back in the word formula again.

And now a word formula, written as a list of instructions, which converts degrees Celsius into degrees Fahrenheit:

Multiply by 9, divide by 5, then add 32

Example

What are these temperatures in Fahrenheit?

(a) 0 °C $\frac{0 \times 9}{5} + 32 = 32\,°\text{F}$

(b) 40 °C $\frac{40 \times 9}{5} + 32 = 104\,°\text{F}$

(c) 100 °C $\frac{100 \times 9}{5} + 32 = 212\,°\text{F}$

Diagrams like the word formula are very useful.

Here are two more.

- Function machine

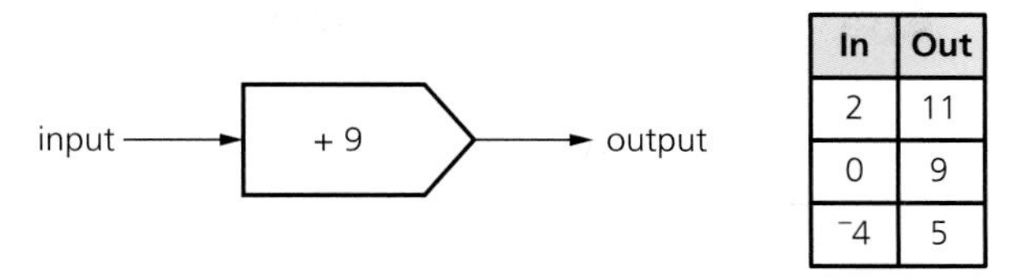

In	Out
2	11
0	9
$^{-}4$	5

This function machine adds 9 to each input.

- Flow chart

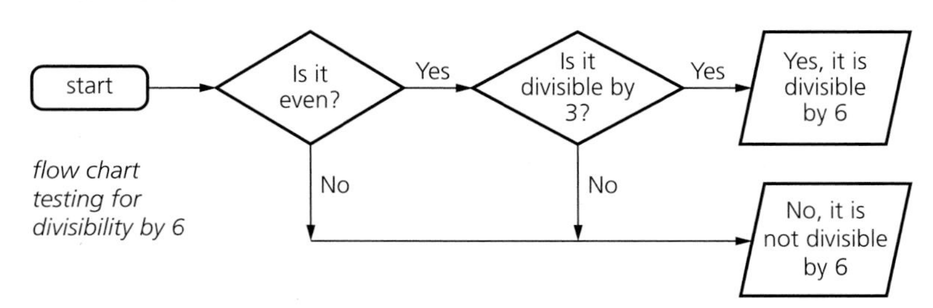

flow chart testing for divisibility by 6

This flow chart is more complex than the previous one as it has 'branches'.

This type of flow chart can be used for sorting inputs and you may have seen a similar flow chart used as an identification key in science.

Exam-style questions

Try these questions for yourself. The answers are given near the back of the book.

Some of the questions involve ideas met in earlier work that may not be covered by the notes in this chapter.

7.1 Find the numbers represented by the **symbols** in these **equations**.

(a) 6 + ▲ = 12 (1)

(b) ■ – 4 = 19 (1)

(c) 6 × □ = 18 (1)

(d) 3 × ☆ × 5 = 15 (1)

(e) 24 + 21 = ◭ × 5 (2)

7.2 Find the unknown number in each case.

(a) Mike is three years older than Emma. He is 12

12 – 3 = ⊙

How old is Emma? (1)

(b) A regular pentagon has a perimeter of 30 cm.

5 × ☆ = 30

What is the length of a side of the pentagon? (1)

7.3 Amy has made a machine that subtracts 4 from every **input** number.

input 5 → – 4 → output 1

(a) What will be the **output** if Amy puts 7 into the machine? (1)

(b) What will be the output if Amy puts 3 into the machine? (2)

(c) If 6 comes out of the machine, which number did Amy put in? (1)

7.4 **(a)** Sally has thought of a number. When she adds 4 she gets 13

What is Sally's number? (1)

(b) Fred has thought of a number. When he subtracts 6 he gets 17

What is Fred's number? (1)

(c) Alice has thought of a number. When she multiplies it by 7 she gets 42

What is Alice's number? (1)

(d) Graham has thought of a number. When he divides it by 9 he gets 6

What is Graham's number? (1)

7.5 Jenny has made a machine.

input 7 → [× 3] → 21 → [− 5] → output 16

Her machine multiplies by 3 and then subtracts 5

(a) What will be the output if Jenny puts in 4? (1)

(b) What will be the output if she puts in 1? (2)

(c) If 1 comes out, which number did Jenny put in? (2)

7.6 **(a)** Bella has thought of a number. When she adds 4 and then multiplies by 3 she gets 18

What is Bella's number? (2)

(b) Colin has thought of a number. When he multiplies by 3 and then subtracts 4 he gets 11

What is Colin's number? (2)

Revision tip

You could draw or make a two-stage function machine with removable labels and challenge a friend to find the labels that will give a particular set of input/output values.

input → [ ]–[ ] → output

7.2 Sequences and functions

Exploring and describing number patterns

It is very useful to be able to spot, explore and describe patterns in a series of numbers.

Look at these arrangements made with matchsticks:

arrangement 1	arrangement 2	arrangement 3
4 sticks	7 sticks	10 sticks

Can you see a pattern here?

How many matchsticks are there in arrangement 4 and arrangement 5?

One of the best ways to explore number patterns is to look at the difference between each number and the next, as in this diagram:

4 ----► 7 ----► 10 ----► ?
+ 3   + 3   + 3

We can then say: 'It is going up in threes' and predict that the next number will be 13

Let us see if we were right.

arrangement 4

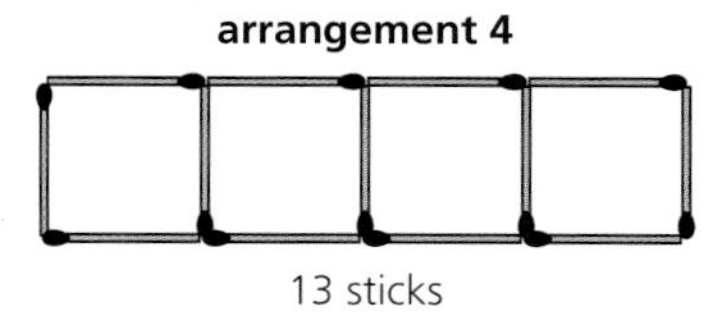

13 sticks

Yes, 13 *is* correct.

Now, can you predict the number of matchsticks needed for arrangement 10?

Revision tip

In a notebook, write down any Maths words in this chapter that are unfamiliar (for example *formula*, *function*, *sequence*, *equation*). Now use the Glossary at the back of the book to help you write the meanings of these terms in your own words. Add other words to your notebook as you go along.

Exam-style questions

Try these questions for yourself. The answers are given near the back of the book.

Some of the questions involve ideas met in earlier work that may not be covered by the notes in this chapter.

7.7 Clare has started to thread beads onto a string in a pattern.

Continue the pattern for the next eight beads. (2)

7.8 Write the next two numbers in each of these sequences.

(a) 2 4 6 8 10 ... (1)

(b) 2 6 10 14 18 ... (2)

(c) 30 25 20 15 10 ... (2)

(d) $\frac{1}{4}$ $\frac{1}{2}$ $\frac{3}{4}$ 1 $1\frac{1}{4}$ ... (2)

7.9 Write the next two numbers in each of these sequences.

(a) 2 6 4 8 6 ... (2)

(b) 1 2 4 7 11 ... (2)

7.10 **(a)** Which multiplication function machine could produce all these outputs?

5 10 45 60 95 1005 (2)

(b) Which multiplication function machine could produce all these outputs?

12 24 30 45 102 3000 (2)

7.3 Graphs

First quadrant co-ordinates

Reading and plotting co-ordinates using positive whole numbers

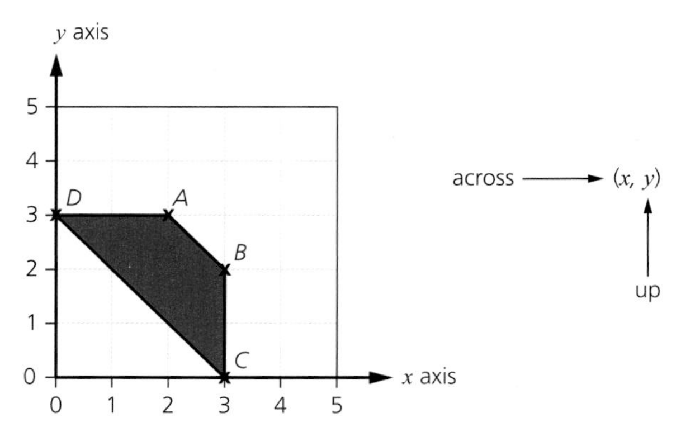

Points A, B, C and D have been plotted on the **grid**.

The point (0, 0) is the reference point for all **co-ordinates**. It is called the **origin**.

The point A is at (2, 3): 2 across, 3 up.

The point B is at (3, 2): 3 across, 2 up.

The point C is at (3, 0): 3 across, 0 up. It stays on the x-**axis.**

The point D is at (0, 3): 0 across, 3 up. It stays on the y-axis.

Shapes drawn on a grid

Points A, B, C and D have been joined in order to form the **shape** $ABCD$.

If we were to slide the shape across or up (or both), we could redraw it anywhere else on the grid. This sliding move is called **translation**. Although the shape would look the same, the co-ordinates of its **vertices** would change.

You also need to be able to answer questions about shapes drawn on a grid.

Example

What shape is *ABCD?*

Isosceles trapezium.

Revision tip

To help with writing the co-ordinates in the correct order, some people think of 'into the house and then up the stairs!'

Exam-style questions

Try these questions for yourself. The answers are given at the back of the book.

Some of the questions involve ideas met in earlier work that may not be covered by the notes in this chapter.

7.11 (a) Complete the table of outputs for this function machine. (2)

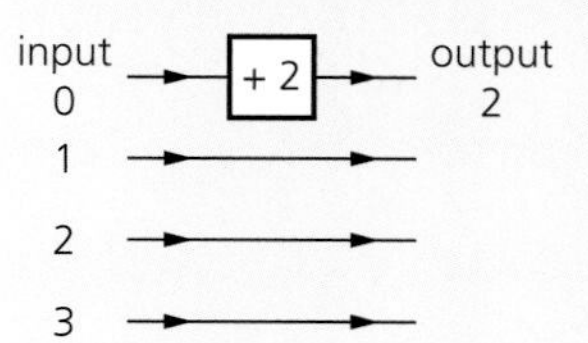

(b) Plot points representing the pairs of input and output numbers on this grid. (2)

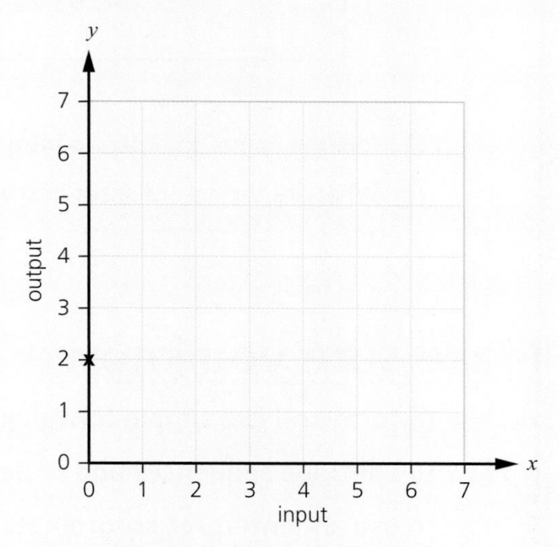

7.12 A function machine has produced the input/output pairs of numbers plotted on this grid.

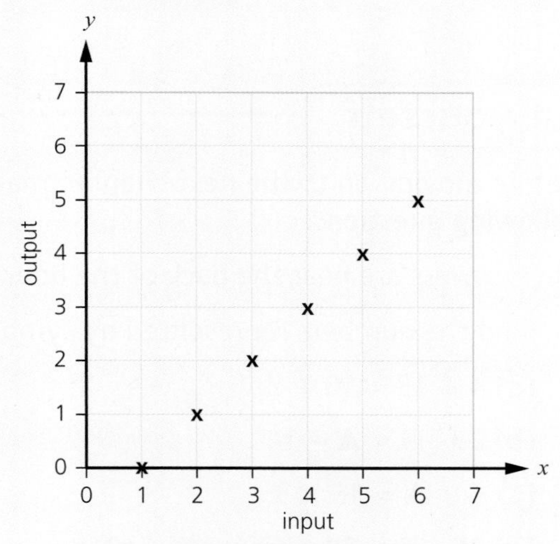

(a) What would be the output for input 7? (1)

(b) What would be the input for output 7? (1)

(c) Complete this word equation which describes the function.

Output = ... (2)

7.13 (a) On this grid, plot the points *A* (3, 0), *B* (6, 1), *C* (7, 6) and *D* (4, 5). (2)

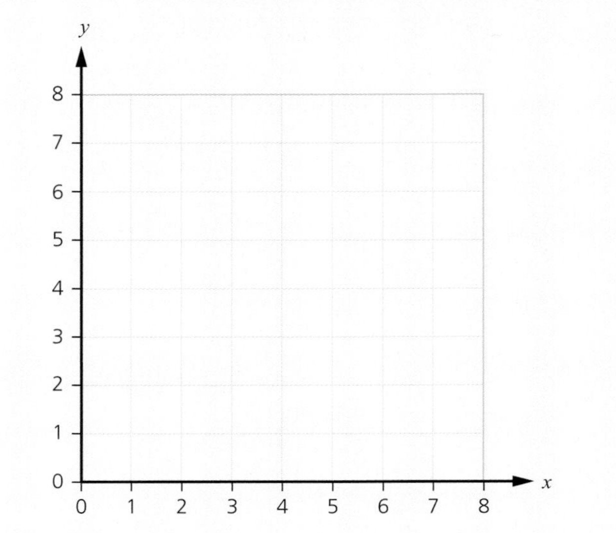

(b) Join the points, in order, to form the shape *ABCD*. (1)

(c) What are the co-ordinates of the mid-point of *ABCD*? (1)

★ Make sure you know

★ How to form and use simple formulae expressed in words or as a picture

★ How to continue sequences and understand simple function machines

★ How to use and interpret co-ordinates in the first quadrant

★ How to use the glossary at the back of the book for definitions of key words

Test yourself ✓

Before moving on to the next chapter, make sure you can answer the following questions.

The answers are near the back of the book.

1 Find the numbers represented by symbols in the following equations.

(a) 4 + 13 = 10 + ☆

(b) 23 – 9 = ▲ – 14

(c) 6 × Δ = 16 × 3

(d) 35 ÷ 5 = 28 ÷ ⊙

2 Will and Gina have each thought of a positive whole number less than 9

The sum of their numbers is 12

We could write this as $w + g = 12$, where w is Will's number and g is Gina's number.

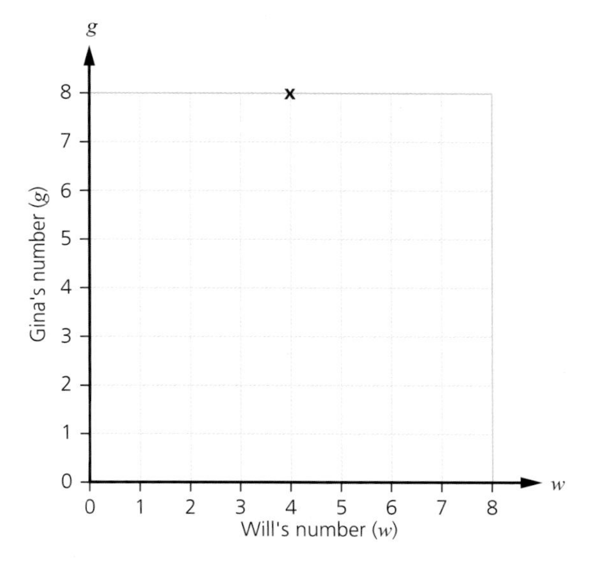

On the diagram, one possibility is shown by a cross.

Complete the diagram to show all of the other possibilities.

3 **(a)** Tommy has thought of a number. When he multiplies it by 3, he gets exactly the same result as he would by adding 8 to his number. What is his number?

(b) Nancy has thought of a number. When she subtracts 3 from twice her number, the result is $1\frac{1}{2}$ times her number. What is her number?

4 Look at these arrangements of matches.

arrangement 1

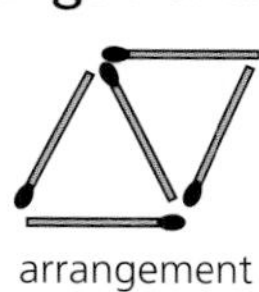

arrangement 2

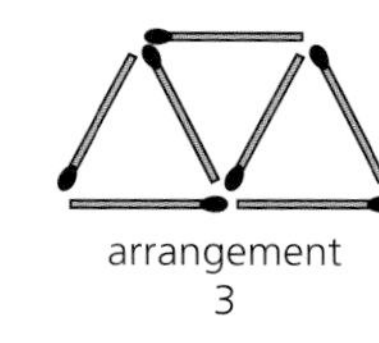

arrangement 3

(a) Sketch arrangements 4 and 5

(b) **(i)** How many matches would be needed for arrangement 6?

(ii) How many matches would be needed for arrangement 10?

(c) Gloria suggests that it is easy to find the number of matches needed. She says you simply multiply the number of triangles by 2 and then add 1

(i) Does Gloria's idea give the correct number of matches for arrangement 6 and arrangement 10?

(ii) Use Gloria's idea to find how many matches would be needed for arrangement 100

5 Here is an unusual arrangement of counting numbers up to 53

0	1	2	3	4	5	6	7	8
9	10	11	12	13	14	15	16	17
18	19	20	21	22	23	24	25	26
27	28	29	30	31	32	33	34	35
36	37	38	39	40	41	42	43	44
45	46	47	48	49	50	51	52	53

(a) Circle all the numbers that would be produced by a × 3 machine.

(b) Shade all the numbers that would be produced by a × 5 machine.

6 **(a) (i)** On this grid, plot the points *A* (3, 4), *B* (5, 6), *C* (3, 8) and *D* (1, 6).

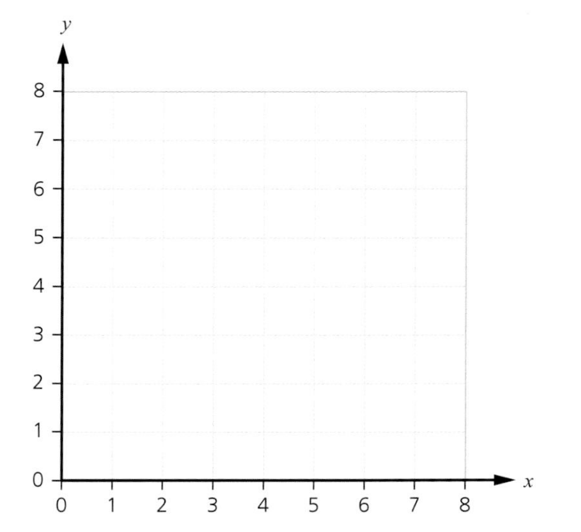

(ii) Join the points in order to form the shape *ABCD* and name *ABCD*.

(b) In a space on the grid, plot three points *P*, *Q* and *R* which when joined make an isosceles triangle. Write down the co-ordinates of *P*, *Q* and *R*.

8 Algebra (2)

8.1 Equations and formulae

Simple symbolic formulae

A quick review of the basics

- We do not use × (times) or ÷ (division) signs.

Example

ab means a times b $\frac{a}{b}$ means a divided by b

- Like terms.

Like terms have exactly the same letter part. These pairs are like terms:

Example

$3a$ and $7a$ $5b$ and ^-2b $6ab$ and ab $3a^2$ and $8a^2$

- Collecting like terms.

Example

$a + 5b - 3ab + 3a + 4ab - b = 4a + 4b + ab$

- Multiplying out brackets.

Remember to multiply the number outside by everything in the brackets. And watch the plus and minus signs!

Examples

$3(a + 2b) = 3a + 6b$

Level 3 $a(b - 2c) = ab - 2ac$

- Factorising into brackets.

Examples

$4a + 4b = 4(a + b)$

Level 3 $15p^2 - 10pq = 5p(3p - 2q)$

- Equivalent fractions.

The numerator and denominator must always be divided (or multiplied) by the same number.

Example

$\frac{4}{12} = \frac{2}{6} = \frac{1}{3} = \frac{10}{30} = \ldots$ etc.

- Simplifying powers.

When multiplying, you *add* the powers; when dividing you *subtract* the powers.

Level 3

Examples

$aa = a^2$

$a(a + a^2) = a^2 + a^3$

$\frac{12a^3}{4a^2} = \frac{3a^3}{a^2} = 3a$

- Letters stand for numbers.

In algebra, a letter always represents a number, not a word. Sometimes we write things that are always true (an **identity**), and sometimes we write things that are only true for one number (an **equation**). A collection of letters and numbers is called an **expression**.

Examples

$a + a = 2a$ is always true, whatever number you use for a (an identity).

$t - 4 = 7$ is only true when $t = 11$ (an equation).

$2a + 3b$ and $5(p - q^2)$ are examples of expressions.

BODMAS and algebraic substitution

Armed with the knowledge of everything above, we can now tackle a wide variety of questions involving algebraic substitution. But first we must find out – who or what is **BODMAS**?

most important (do first) ↑	**B**rackets	
	Of (e.g. square or square root of)	
	Division	equal priority; work from left to right
	Multiplication	
	Addition	equal priority; work from left to right
least important (do last) ↓	**S**ubtraction	

Some people prefer the acronym **BIDMAS** – where the I stands for **Indices** (powers). The main thing to remember, though, is that multiplication/division must always be done before addition/subtraction.

BODMAS means that there is now no doubt about the answer to ambiguous questions: $3 + 4 \times 5 = 23$ (not 35) because we work out the 4×5 first.

Revision tip

Your calculator may be a BODMAS calculator. Type in one of the examples in this section to check! There are two types of calculator: LTR and MDF. Type in $10 - 2 \times 3$ and then press equals. LTR (Left To Right) gives the answer 24, while MDF (Multiplication and Division First) gives 4. MDF knows BODMAS. Does yours? Know your calculator before you go into any exam!

Algebraic examples

(a) Given that $p = 2q + 3(q + r)$, find p when $q = 3, r = 4$

First note that $2q$ is 6, and that $(q + r)$ is 7

Then: $p = 6 + 3 \times 7$

$p = 27$ [since $3 \times 7 = 21$]

(b) Given that $E = mc^2$, find E when $m = 10$ and $c = 5$

$E = 10 \times 5^2$

$E = 10 \times 25$

$E = 250$

Level 1 and 2
Papers may include examples of algebraic fractions such as $\frac{3x}{15}$

Level 3
Papers may include examples of algebraic fractions where the variable appears in both the numerator and the denominator, for example $\frac{4x}{x}$

Solving simple linear equations

Revision tip
Many people find it helpful to think of equations as a set of balance scales, with the = sign as the pivot. At every stage of your working, be careful not to upset the balance!

How to solve equations: some guidelines

- At every stage apply the same action to both sides of the equation.
- Think about the order in which the equation was written (BODMAS may help you), and work backwards to the unknown variable by *undoing* each step in reverse order.
- Try to collect letter terms on one side and number terms on the other.
- Don't be put off if the answer is negative or a fraction.

Example 1

Solve the equation $x - 7 = 20$

Equation	Do this to both sides
$x - 7 = 20$	Add 7
$x = 27$	Check original equation
Check LHS: $27 - 7 = 20$	
Check RHS: $= 20$	It is the right answer!

Example 2

Solve the equation $3x + 15 = 9$

Equation	Do this to both sides
$3x + 15 = 9$	Subtract 15
$3x = {}^{-}6$	Divide by 3
$x = {}^{-}2$	Check original equation
Check LHS: $3({}^{-}2) + 15 = 9$	
Check RHS: $= 9$	It is the right answer!

Example 3

Solve the equation $4x + 5 = 26 - 3x$

Equation	Do this to both sides
$4x + 5 = 26 - 3x$	Subtract 5
$4x = 21 - 3x$	Add $3x$
$7x = 21$	Divide by 7
$x = 3$	Check original equation
Check LHS: $4(3) + 5 = 17$	
Check RHS: $26 - 3(3) = 17$	It is the right answer!

Example 4

Solve the equation $5(19 - x) = 3(3x - 1)$

Equation	Do this to both sides
$5(19 - x) = 3(3x - 1)$	Multiply out the brackets
$95 - 5x = 9x - 3$	Add 3
$98 - 5x = 9x$	Add $5x$
$98 = 14x$	Divide by 14
$7 = x$	Check original equation
Check LHS: $5(19 - 7) = 60$	
Check RHS: $3(21 - 1) = 60$	It is the right answer!

Papers may include simple examples of linear equations such as:

Level 1 ◆ $x - 4 = 10$ $4x = 20$ $2x + 3 = 9$

and slightly more complicated examples such as:

Level 2 ● $6 = 2x - 2$ $2x + 5 = 11 - x$ $2(x + 3) = 12$ $\frac{1}{2}x = 3$

Papers may also include examples such as:

Level 3 ■ $6 - 4x = 3$ $\frac{1}{2}x + 3 = 5$ $\frac{2}{3}(2x - 3) = 12$

Forming and solving linear equations

Here is a difficult puzzle that can be solved more easily by setting up a simple equation.

I have x sweets. You have three fewer than twice the number I have. One-third of your sweets is five fewer than the number I have. How many do we each have?

We make the equation up step by step:

I have x sweets.

You have three fewer than twice the number I have.

So you have three fewer than twice x.

You have $2x - 3$ sweets.

One-third of your sweets is five fewer than my number of sweets.

So one third of $(2x - 3)$ is $x - 5$

So the equation is $\frac{1}{3}(2x - 3) = x - 5$

Now we can solve the equation.

$\frac{1}{3}(2x-3) = x - 5$	Multiply both sides by 3
$2x - 3 = 3x - 15$	Add 15 to both sides
$2x + 12 = 3x$	Subtract $2x$ from both sides
$12 = x$	So I had 12 sweets ...
$2(12) - 3 = 21$	... and you had 21

Level 3

Solving quadratic equations

A **root** of an equation is a value of x that will make the value of that equation equal to zero. Quadratic equations can have two different roots.

Using trial and improvement methods

Trial and improvement helps you to zoom in on one of those roots. You are usually told where to start looking and then each subsequent trial improves your answer.

Example

The equation $x^2 - 3x + 1 = 0$ has a root (solution) between $x = 2$ and $x = 3$

Find this value of x to two decimal places.

x	$x^2 - 3x + 1$	Verdict
2	$^-1$	Too small
3	1	Too big
2.5	$^-0.25$	Too small
2.7	0.19	Too big
2.6	$^-0.04$	Too small
2.65	0.0725	Too big
2.62	0.0044	Too big
2.61	$^-0.0179$	Too small
2.615	$^-0.006\,775$	Too small

We always choose our next x value to be half way between the *most* recent *too small* and *too big* values.

After all our working, our conclusion is that x must be between 2.615 and 2.62 Both of these x values, when expressed to two decimal places, give 2.62

Using the calculator TABLE function

If your calculator has a TABLE function, then you may set out this table in a different way.

First, set the step size to 0.1 (between 2 and 3) and notice that you get a change of sign between 2.6 and 2.7 (write these values in your answer).

Next, set the step size to 0.01 (between 2.6 and 2.7) and notice that you get a change of sign between 2.61 and 2.62 (write these values in your answer).

Finally set the step size to 0.001 (between 2.61 and 2.62) and notice that you get a change of sign between 2.618 and 2.619, each of which rounds to 2.62 to 2 d.p, so this is your final answer.

In order to give an answer to two decimal places, we must calculate to three decimal places.

If x is between two numbers which both round to 2.62 to 2 decimal places, then x must also round to 2.62 to 2 decimal places.

Level 3

Solving inequalities and finding integer solution sets

Inequalities

Consider the inequality: $5 + \frac{1}{2}x > 4$

Subtract 5 from both sides $\frac{1}{2}x > {}^{-}1$

Multiply by 2 $x > {}^{-}2$

So x could be ⁻1, 0, 1, 2, 3, 4, 5, ...

Any of these would make $\left(5+\frac{1}{2}x\right)$ more than 4

We could show the solution on a number line, by shading out the part that we do not want:

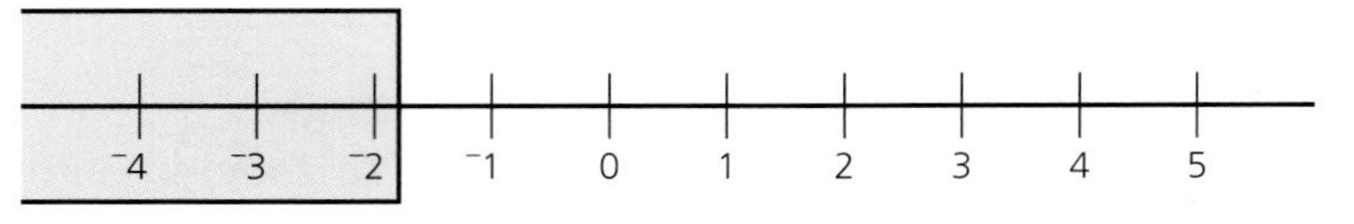

Considering the example above,
suppose further that $2x - 5 < 1$

Add 5 to both sides $2x < 6$

Divide through by 2 $x < 3$

Represented on the number line as:

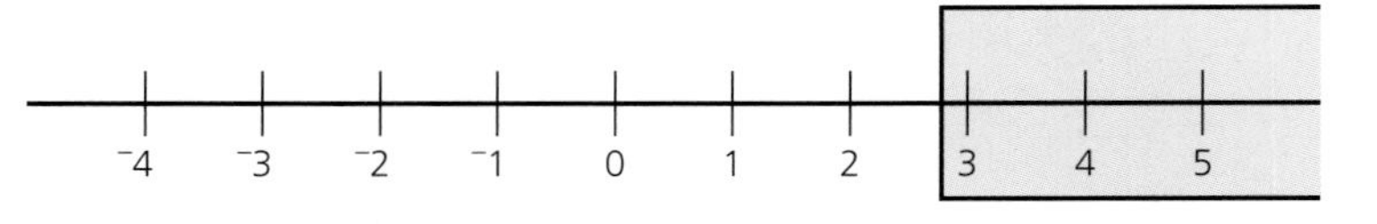

Then the set of integers which satisfies both inequalities at the same time (the integer solution set) is shown by this combination diagram:

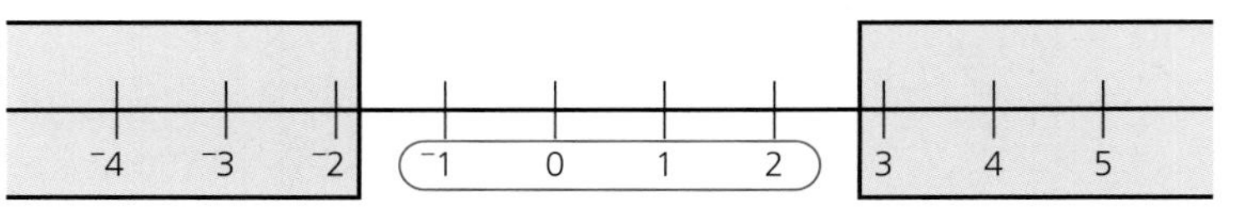

Not examined
This section is included for interest only, and is unlikely to be tested in a Common Entrance examination or Scholarship.

Algebraic formulae

Multiplying out brackets

$a(x + y) = ax + ay$

$a(x - y) = ax - ay$

$(a + b)(x + y) = ax + ay + bx + by$

$(a + b)(x - y) = ax - ay + bx - by$

$(a - b)(x + y) = ax + ay - bx - by$

$(a - b)(x - y) = ax - ay - bx + by$

Revision tip

There will be tears if you get the signs wrong when multiplying out brackets! You must remember that any product is negative *only* if you are multiplying a positive with a negative in either order. If you are unsure you should substitute values for a, b, x and y then check that the left and right sides give the same result.

Squaring a bracket

$$(a + b)^2 = a^2 + 2ab + b^2$$

$$(a - b)^2 = a^2 - 2ab + b^2$$

$$(a + b + c)^2 = a^2 + ab + ac + ab + b^2 + bc + ac + bc + c^2$$

$$= a^2 + b^2 + c^2 + 2ab + 2ac + 2bc$$

Difference of two squares

$$a^2 - b^2 = (a - b)(a + b)$$

Difference of two cubes

$$a^3 - b^3 = (a - b)(a^2 + ab + b^2)$$

Alphabetical order of letters

When we write an algebraic term, it is conventional to write the letters in alphabetical order, with numbers first. This helps us to spot like terms more easily.

So cad is written as acd, and $f \times 3 \times d^2$ is written as $3d^2f$ and so on.

Pythagoras' theorem

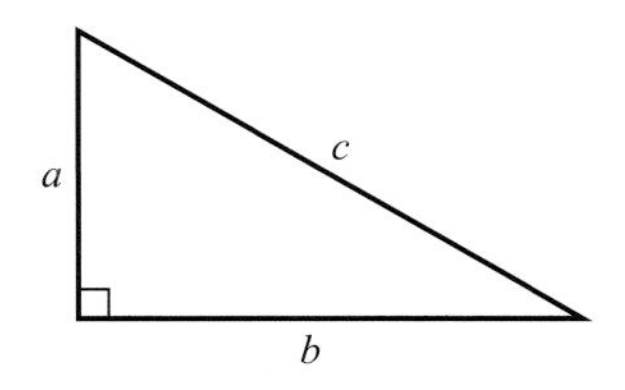

$$a^2 + b^2 = c^2 \qquad c = \sqrt{a^2 + b^2}$$

$$b^2 = c^2 - a^2 \qquad b = \sqrt{c^2 - a^2}$$

$$a^2 = c^2 - b^2 \qquad a = \sqrt{c^2 - b^2}$$

Examples

These are examples of Pythagorean triples (triangles with integer side lengths):

(3, 4, 5) (5, 12, 13) (6, 8, 10) (9, 12, 15) (10, 24, 26) ...

Exam-style questions

Try these questions for yourself. The answers are given near the back of the book.

Some of the questions involve ideas met in earlier work that may not be covered by the notes in this chapter.

8.1 Simplify these expressions:

(a) $5c - 3c + 6c$ (1)

(b) $4d^2 + 2d^2 - 5d^2$ (2)

(c) $5e^2 \times 3e$ (2)

(d) $\dfrac{9f}{3}$ (2)

Level 3 ■ (e) $\dfrac{9b+3b}{6b}$ (2)

8.2 Factorise completely:

(a) $3h - 12$ (2)

(b) $20e - 15f$ (2)

Level 3

■ (c) $2e^2 + 4e$ (2)

■ (d) $4p^2 + 8pq$ (2)

8.3 Multiply out the brackets and simplify:

(a) $4(a + 3b)$ (1)

(b) $2(3q + 3r) - 2r$ (2)

(c) $3(m + 2n) - (m + 3n)$ (3)

8.4 Given that $d = 4$, $e = 6$ and $f = 5$, find the value of:

(a) $d - e$ (2)

(b) def (2)

(c) $\frac{d+e}{f}$ (2)

(d) $d^2 - f^2$ (3)

8.5 If $m = 3$, $n = {}^{-}2$ and, $p = 4$, find the value of each of these expressions:

(a) $m + n$ (1)

(b) np (1)

(c) $m(n - p)$ (2)

(d) mn^2 (2)

8.6 Tom has thought of a number which he has called t.

Write an expression for:

(a) the number that is 3 more than Tom's number (1)

(b) the number that is 4 times Tom's number (1)

(c) the number that is 5 less than Tom's number (1)

(d) the number that is half of Tom's number (1)

(e) the square of Tom's number. (1)

8.7 In a game of numbers, Rosie chooses a number b.

Ellie chooses a number that is two more than Rosie's number.

(a) Write down an expression, in terms of b, for Ellie's number. (1)

Kori chooses a number which is three times Rosie's number.

(b) Write down an expression, in terms of b, for Kori's number. (1)

(c) Write down an expression, in terms of b, for the sum of the three numbers. Simplify your answer. (2)

The sum of the three numbers is 22

(d) Form an equation and solve it to find the number that Rosie chose. (2)

8.8 A pork pie costs x pence.

(a) Write down, in terms of x, the cost of five pork pies. (1)

A cup of tea costs 3 pence more than a pork pie.

(b) Write down, in terms of x, the cost of a cup of tea. (1)

(c) Write down, in terms of x, the cost of seven cups of tea. (2)

(d) Write down, in terms of x, the cost of three pork pies and four cups of tea. Simplify your answer. (2)

Mandy buys three pork pies and four cups of tea for £4.67

(e) Form an equation, in terms of x, and solve it to find the cost in pence of a pork pie. (3)

(f) If Mandy had bought only two pork pies and two cups of tea, how much change should she have received from a £10 note? (2)

8.9 Solve these equations:

(a) $q + 8 = 12$ (1)

(b) $3p = 12$ (1)

(c) $4n - 3 = 9$ (2)

(d) $3y + 7 = 13$ (2)

(e) $6x + 5 = 4x - 10$ (3)

8.10 Solve these equations:

(a) $4s = s + 12$ (2)

(b) $4a - 5 = {}^{-}3$ (2)

(c) $5 - 2r = 2$ (2)

(d) $5(y + 3) - 4(y - 5) = 50$ (3)

(e) $\frac{1}{2}m = 13$ (2)

8.11 Solve these equations:

Level 2

● (a) $21 - 2a = 7$ (2)

● (b) $3(u - 2) = 4(2 - u)$ (3)

Level 3

■ (c) $4 - 2(3v + 5) = 0$ (3)

■ (d) $\frac{1}{2}c - 6 = 8$ (2)

■ (e) $\frac{3}{4}(v + 4) = 5$ (3)

■ (f) $3(s - 4) = 16 - s$ (3)

■ **8.12** For the function $y = x^2 - 2x$ find, by 'trial and improvement', the positive value of x when y is 30

Complete a table like the one below, or use the TABLE function on your calculator, giving your final answer to three significant figures. (5)

x	x^2	$2x$	$x^2 - 2x$
6			
7			

■ **8.13** (a) (i) Solve the inequality $4 + 5x < 15$ (2)

(ii) List the positive integers which satisfy the inequality in part (i). (1)

(b) (i) Solve the inequality $6 - \frac{1}{2}n \geq 3$ (3)

(ii) Write down the largest value n can take in part (b) (i). (1)

8.2 Sequences and functions

Describing in words the next term of a sequence

As a warm-up to this section, write down the next two terms of each of these sequences:

(a) 3, 5, 7, 9, 11, 13, ..., ..., 15, 17

(b) 27, 23, 19, 15, 11, 7, ..., ..., 3, ⁻1

(c) 2, 3, 5, 8, 12, 17, ..., ..., 23, 30

(d) 32, 16, 8, 4, 2, 1, ..., ..., 0.5, 0.25

'Describing in words' is when you explain what the sequence is doing. From the previous sequences:

(a) adding 2 each time

(b) subtracting 4 each time

(c) adding one more than the previous time

(d) dividing by 2 each time.

Remind yourself again of the method described in Chapter 7, Section 7.2.

This method (the **difference method**) will show up almost every hidden pattern.

Describing the *n*th term of a linear sequence

Describing the *n*th term of a sequence in words

Example

Simple sequences look very much like the answers in the times tables:

4, 8, 12, 16, ...

To find the next term you either add 4 to the previous one, or work it out directly using the four times table. So to find the 5th term, you just work out $5 \times 4 = 20$

It is all very well being able to find the next term in a sequence, but could you find the 10th term, the 100th term, without plodding through all the other terms along the way?

If a sequence is going up or down by the same amount each time, then that number becomes your multiplying number, and you just need to adjust it by adding or subtracting another number.

Example

Find the 100th term of the sequence: 2, 5, 8, 11, 14, ...

Note that this is going up each time by 3

first term:	2	$1 \times 3 = 3$
second term:	5	$2 \times 3 = 6$
third term:	8	$3 \times 3 = 9$
fourth term:	11	$4 \times 3 = 12$

Look at the last column: 3, 6, 9, 12 – it is always one more than the number we want (2, 5, 8, 11).

Multiplying by 3 is therefore not enough. We then have to subtract 1

Fifth term should be: 14

Find it by calculation. **5** × 3 = 15

15 – 1 = **14** Yes! It works!

Now we can find the 100th term easily.

Hundredth term: **100** × 3 = 300

300 – 1 = **299**

Level 3

Describing the *n*th term of a linear sequence algebraically

This follows on very simply and neatly from the last section. We just need to learn a bit of special notation for sequences.

n The number of the term indicates the **position** in the sequence:

$n = 1$ for the first term in the sequence.

$n = 2$ for the second term, and so on.

T_n As for the nth term itself, T stands for *term*, and n can be any number (as earlier).

Thus we have:

T_1 for the first term

T_2 for the second term and so on.

Example

Find the nth term of the sequence 2, 7, 12, 17, ...

We start by making a table:

n	0	1	2	3	4	5	6	7
T_n	$^-3$	2	7	12	17	22	27	32
increase	5	5	5	5	5	5	5	

The shaded boxes have just been worked out.

What is the *increase* each time?

What would the term *before* the first one be?

The nth term is now obvious! $T_n = 5n - 3$

Level 3

Quadratic sequences: the *n*th term and substitution

Can you find the next two terms of this sequence: 4, 7, 12, 19, 28, ...

We can make a table similar to the one in the example.

n	1	2	3	4	5	6	7
T_n	4	7	12	19	28	39	52
Increase	3	5	7	9	11	13	
Increase		2	2	2	2	2	

As you can see, for quadratic rules we need *two* rows of differences in the table. It turns out (you wouldn't be expected to find it for yourself at Common Entrance or Key Stage 3) that the quadratic rule for the above sequence is

$T_n = n^2 + 3$

It is much easier if you are told the rule at the start. To find each term, you just substitute n = 1, 2, 3, ... into the formula.

Example

Give the first three terms of the sequence $T_n = 2n^2 - 1$

Substituting in $n = 1$, we obtain $2 \times 1 - 1 = 1$

$n = 2$, $2 \times 4 - 1 = 7$

$n = 3$, $2 \times 9 - 1 = 17$

Exam-style questions

Try these questions for yourself. The answers are given at the back of the book.

Some of the questions involve ideas met in earlier work that may not be covered by the notes in this chapter.

8.14 Write down the next two **terms** in each of the following **sequences**:

(a) 47 43 39 35 ... ... (2)

(b) 1 3 7 13 ... ... (2)

(c) 800 400 200 100 ... ... (2)

(d) 3 8 13 18 ... ... (2)

(e) 1 4 16 64 ... ... (2)

(f) 31 15 7 3 ... ... (2)

8.15 (a) A sequence starts with 1

Using the rule 'multiply by 2 and then add 4', write down the next four terms of the sequence. (2)

(b) Write a simple rule for finding the next term in this sequence.

1 4 7 10 13 ... (1)

8.16 Ann has been studying this sequence.

2 5 8 11 14 ...

She has written the formula $T_n = 3n - 1$ for finding the nth term of the sequence.

(a) Use Ann's formula to find the 100th term of the sequence. (1)

(b) Find the smallest value of n for which T_n is more than 1000 (2)

8.17 For the linear function $y = x + 4$, complete this table of input (x) and output (y) values. (2)

x	$^-1$	0	1	2	3	4
y	3	4				

Level 3

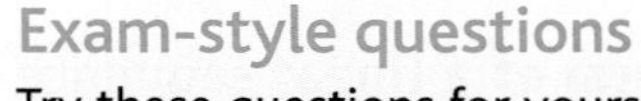

8.18 For the quadratic function $y = x^2$, complete this table of x and y values. (2)

x	$^-2$	$^-1$	0	1	2	3
y	4				4	

8.3 Graphs

Four-quadrant co-ordinates

The four quadrants are the four regions on the grid formed by extending both the x and y axes into the negative numbers.

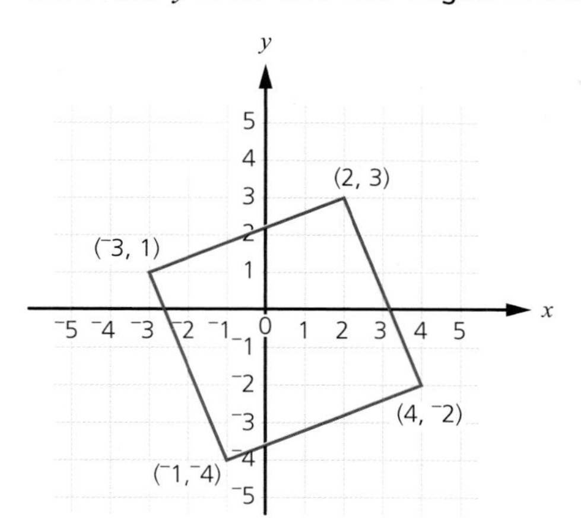

Is this shape

(a) a rhombus

(b) a square

(c) a parallelogram

(d) a trapezium

(e) a kite

(f) a rectangle?

Answer: It is all of them!

Mappings

Graph **mappings**, or line equations, are usually given in one of the forms shown in this table.

	Type 1	Type 2	Type 3	Type 4
Form	$x = k$	$y = k$	$y = mx$	$y = mx + c$
Example	$x = 5$	$y = {}^-3$	$y = 2x$	$y = 2x + 5$

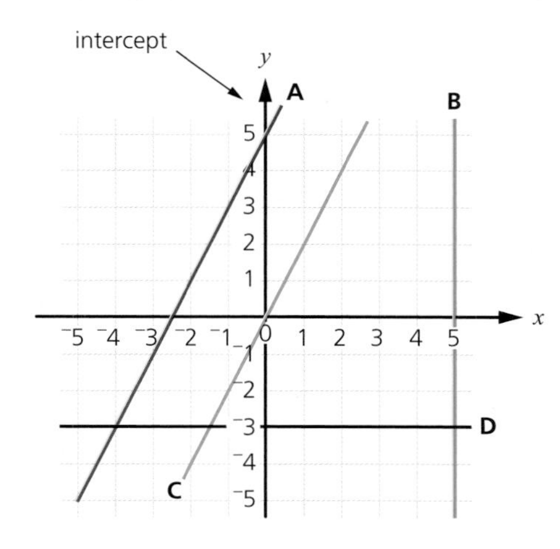

Type 1 are always vertical lines.

Type 2 are always horizontal lines.

Type 3 are always sloping lines going through the origin.

Type 4 always cross both axes.

Can you match each of the lines A, B, C and D with the correct equation from the examples in the table?

Answer: A is type 4, B is type 1, C is type 3 and D is type 2

Note: Papers may test the simple types 1, 2 and 3,

Level 1 ◆ for example $x = 4$, $y = {}^-2$, and $y = x$

and the more complicated types 3 and 4,

Level 2 ● for example $y = 3x$ and $y = 2x - 1$

Papers may also involve mappings of quadratic functions,

Level 3 ■ for example $y = x^2 - 4$

Level 3

Solving simultaneous equations algebraically and graphically

Any pair of simultaneous equations may be solved using algebra or by drawing graphs. Consider the following pair of simultaneous equations:

$2x - 3y = 4$ ①

$3x + 2y = 19$ ②

These equations will be solved both ways to show and compare the two methods in the next section.

Algebraic method

We note that equations ① and ② have opposite signs in the middle. We will therefore want to add them. (If the signs were the same, then we would subtract.) To do this, first make the y **coefficients** the same:

① × 2 is $4x - 6y = 8$ ③

② × 3 is $9x + 6y = 57$ ④

Add ③ and ④ $13x = 65$

$x = 5$

Substitute $x = 5$ into ②:

$15 + 2y = 19$

$2y = 4$

$y = 2$

Check

① $10 - 6 = 4$ OK!

② $15 + 4 = 19$ OK!

Graphical method

We can make a table for each equation so that we can draw them both as straight line graphs.

① becomes $3y = 2x - 4$

x	2	5	8
$2x - 4$	0	6	12
y	0	2	4

② becomes $2y = 19 - 3x$

x	1	3	5
$19 - 3x$	16	10	4
y	8	5	2

which we then plot and draw, as has been done here:

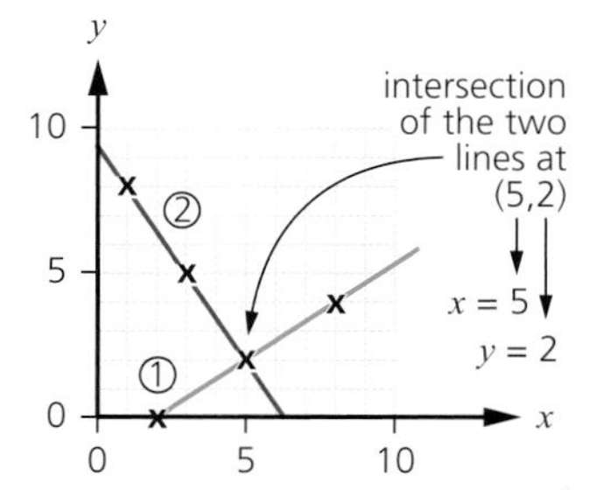

Thus we are able to read off from the intersection of the graphs that the solution to our simultaneous equations is $x = 5$, $y = 2$

Exam-style questions

Try these questions for yourself. The answers are given near the back of the book.

Some of the questions involve ideas met in earlier work that may not be covered by the notes in this chapter.

8.19 Prepare a co-ordinate grid with axes numbered $^{-}5$ to 5. On the grid, draw and label lines representing these functions:

(a) $x = 4$ (1) **(b)** $y = 5$ (1) **(c)** $x = {}^{-}3$ (1)

Level 2

● **8.20 (a)** Tom has prepared this table of x and y values for the function $y = 2x + 3$

x	$^{-}1$	0	1	2	3	4
y	1	3	5	7	9	11

Plot several of the co-ordinate pairs (x, y) on a suitable co-ordinate grid and draw the graph of $y = 2x + 3$ (2)

(b) On the same grid, draw and label the graph of the function $y = x - 4$ (2)

(c) Write down the co-ordinates of the point where the two lines intersect. (1)

Level 3

■ **8.21** Sarah has prepared this table of x and y values for the function $y = x^2 - 2$

x	$^{-}3$	$^{-}2$	$^{-}1$	0	1	2	3
y	7	2	$^{-}1$	$^{-}2$	$^{-}1$	2	7

Prepare a co-ordinate grid with the x-axis numbered from $^{-}3$ to 3 and the y-axis numbered from $^{-}3$ to 8

Plot the co-ordinate pairs (x, y) on the grid and draw the graph of $y = x^2 - 2$ (2)

■ **8.22 (a)** Complete the table of values for the graph of $y = x^2 + 2x$ (3)

x	$^{-}4$	$^{-}3$	$^{-}2$	$^{-}1$	0	1	2
x^2	16		4	1	0		
$2x$	$^{-}8$			$^{-}2$	0		4
y	8			$^{-}1$	0		

(b) Draw the graph of $y = x^2 + 2x$ on a suitable co-ordinate grid. (3)

(c) Use your graph to find the values of x where the graph crosses the line $y = 4$ (2)

(d) Write down the value of x when y takes its lowest value on the graph. (1)

● **8.23** The graphs of two functions, $y = x + 2$ and $y = 6 - x$, are drawn on the grid.

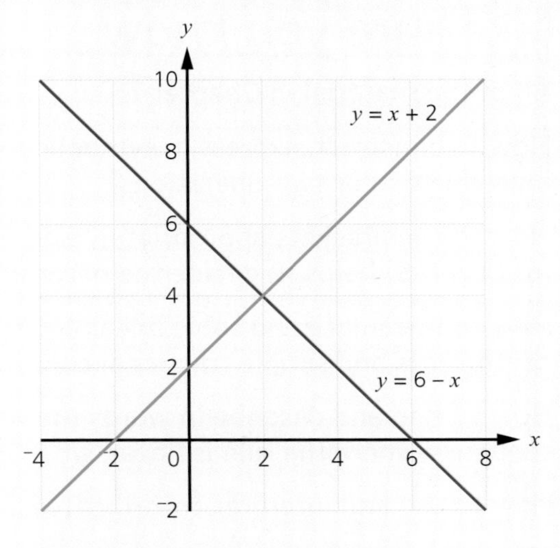

(a) What are the co-ordinates of the point of intersection of the two graphs? (1)

(b) By substitution, check that the co-ordinate pair satisfies both equations. (2)

Level 3

8.24 Consider this pair of simultaneous equations.

$y = 2x - 3$

$y = x + 1$

(a) Substitute y to combine this pair of equations into a single equation in x. (1)

(b) Solve the equation to find x. (2)

(c) Substitute the value of x in one original equation to find the value of y. (1)

(d) Check your answers by substitution in the second equation. (1)

8.25 Consider this pair of simultaneous equations.

$3y = x + 5$

$y = {}^{-}x + 3$

(a) Eliminate x by adding the two equations, thus forming an equation in y. (1)

(b) Solve the equation to find y. (1)

(c) Substitute the value of y in one original equation to find the value of x. (2)

8.26 Apples cost x pence each and pears cost y pence each.

(a) Three apples and two pears cost £1.10

Write this information as an equation in terms of x and y. (1)

(b) Five apples cost the same as four pears.

Write this information as an equation in terms of x and y. (1)

(c) Use your answers to parts (a) and (b) to find the values of x and y. (4)

(d) Tamara spends exactly £2.25 when buying a selection of apples and pears for her friends.

Write down all the different combinations of apples and pears that she could buy. (3)

★ Make sure you know

- ★ All of the material in Chapter 7
- ★ How to construct, express in symbolic form, and use simple formulae involving one or two operations
- ★ How to use brackets appropriately and understand how BODMAS (or BIDMAS) helps to decide the order of operations
- ★ How to formulate and solve linear equations with whole-number coefficients and simple fractional coefficients
- ★ How to find and describe in words the rule for the next term of a sequence where the rule is linear, when exploring number sequences
- ★ How to use and interpret co-ordinates in all four quadrants
- ★ How to represent mappings expressed algebraically, and use Cartesian co-ordinates for graphical representation, interpreting general features

- ★ How to represent mappings of simple linear functions

Level 3
- ■ ★ How to find the nth term of a linear sequence
- ■ ★ How to represent mappings of more complex linear functions
- ■ ★ How to solve quadratic equations, using trial-and-improvement methods
- ■ ★ How to solve simple inequalities
- ■ ★ How to describe the rule for a linear sequence algebraically
- ■ ★ How to find and describe in symbols the next term or the nth term of a sequence where the rule is quadratic
- ■ ★ How to represent mappings of quadratic functions
- ■ ★ How to use algebraic and graphical methods to solve simultaneous linear equations in two variables
- ★ How to use the glossary at the back of the book for definitions of key words

Test yourself

Before moving on to the next chapter, make sure you can answer the following questions. The answers are near the back of the book.

Level 3

1 Simplify:

(a) $3k + 3k$

(b) $5k - 2k + 3k$

(c) $k \times 3k$

(d) $(3k)^2$

(e) $6c^3 + 3c^3$

(f) $4t^3 \times 5t^3$

2 Multiply out the brackets and simplify, where possible:

(a) $4(a - 2)$

(b) $3(2a + 5)$

(c) $2(3a - 4) + 5a - 2$

(d) $5(2t + 3u) - 2v$

(e) $5e - 2(2e + 3f)$

3 Simplify:

(a) $5rs + r^2 - 3rs + 4r^2$

(b) $(2b)^2 - 4b^2$

(c) $4(5b + 2) - 5(3b - 3)$

■ **(d)** $4d^3 \div 5cd^3$

■ **(e)** $\dfrac{9b^2 + 6b^2}{3b}$

■ **(f)** $\dfrac{4n^2 + (2n)^2}{n}$

4 Factorise completely:

(a) $3a + 6$

(b) $4b - 10$

(c) $2a + 4b - 8$

■ **(d)** $a^2 + 2a$

■ **(e)** $6r^3 - 3r^2$

5 When $a = 5$, $b = {}^-2$ and $c = 6$, find the value of:

(a) $2a - b$

(b) $2b^2$

(c) $c(a + b)$

(d) $\dfrac{2ac}{b}$

6 Sam is four years older than Jack, who is c years old.

(a) How old, in terms of c, is Sam?

Six years ago, Sam was twice as old as Jack was.

(b) Form an equation and solve it to find the value of c

(c) How old will Sam be in three years' time?

7 y is an integer.

(a) Write down, in terms of y, the integer that is

(i) four less than y

(ii) three times y.

(b) Calculate, in terms of y, the mean of the three integers.

(c) The mean is 22; form an equation in y and solve it.

(d) Use your answer to part (c) to write down the three integers.

8 Solve the following equations:

(a) $d - 7 = 12$

(b) $4w + 3 = 15$

(c) $4 + 4c = 16$

(d) $5a + 6 = 1$

(e) $5e + 3 = 2e + 9$

(f) $b - 9 = {}^{-}3$

9 Solve the following equations:

(a) $\frac{1}{4}d = 3$

(b) $\frac{3}{4}r = 15$

(c) $2y + 4 = 4y - 12$

Level 3

■ **(d)** $5 - 4x = 11$

■ **(e)** $\frac{1}{4}x - 8 = {}^{-}2$

■ **(f)** $\frac{1}{4}(x - 8) = {}^{-}2$

■ **10** The function $y = 12x - x^2$ can be written in the form $y = x(12 - x)$.

(a) By trial and improvement find, correct to 3 significant figures, the value of x when y is 30

Complete a table like this one.

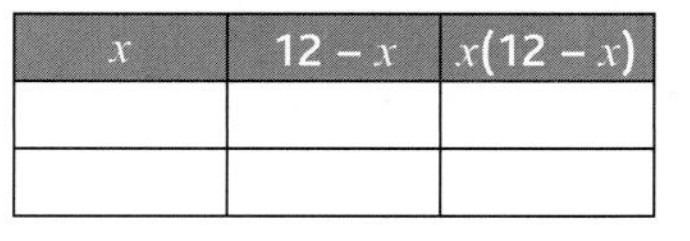

x	$12 - x$	$x(12 - x)$

When y is 32, there are two possible values for x:

x could be 8 since $8(12 - 8) = 32$

x could be 4 since $4(12 - 4) = 32$

(b) Find a second value for x when y is 30

(c) Find two values for x when y is 20

11 Write down the next two terms in each of the following sequences:

(a) 21 18 15 12 ...

(b) 2560 640 160 40 ...

(c) 1 3 4 7 11 ...

12 Find the 6th term and the 8th term of each of the following sequences:

(a) 3 6 9 12 15 ...

(b) 1 4 9 16 25 ...

(c) 1 2 5 10 17 ...

13 **(a)** Apply the rule 'multiply by 3 and then subtract 2' to write down the next four terms of the sequence beginning 2, 4, ...

(b) Apply the rule 'subtract 2 and then multiply by 3' to write down the next four terms of the sequence beginning 4, 6, ...

14 **(a)** The rule 'multiply by 2 and then add 1', starting with 1, produces this sequence:

1 3 7 15 31 ...

Suggest a similar rule which will give this sequence:

2 3 5 9 17 ...

(b) Suggest a formula of the form $T_n = \ldots$ which represents this sequence:

3 7 11 15 19 23 ...

Level 3

■ **15** **(a)** **(i)** Write down the first term (T_1) and 100th term (T_{100}) of the sequence that can be written as $T_n = n^2 - 2$

(ii) Find the smallest value of n for which $T_n > 400$

(b) **(i)** Write down the first and 100th terms of the sequence that can be represented by $T_n = \frac{2n-1}{3n+1}$

(ii) What happens to T_n as n gets very large?

■ **16** **(a)** Complete the table of values for the function $y = 2x^2 + x - 2$

x	$^-3$	$^-2$	$^-1$	0	1	2	3
x^2	9						
$2x^2$	18						
y	13						

(b) On a suitable co-ordinate grid, plot the (x, y) pairs and sketch the curve for $^-3 \leq x \leq 3$

Level 3

■ **17** Molly is m years old and her younger sister Natascha is n years old.

(a) Write down expressions, in terms of m and n, for

(i) the sum of their ages

(ii) the difference between their ages.

They calculate that the sum of their ages is 20 years.

(b) Write down an equation, in terms of m and n, to show this information.

Molly is 6 years older than Natascha.

(c) Use this information and your answer to part (b) to find Natascha's age.

Geometry and measures (1)

9.1 Measures

Reading scales

There are many common scales used for measuring so take care when reading them.

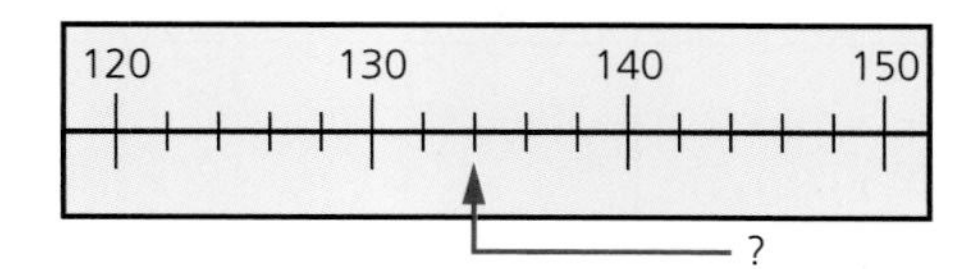

In this first example, we can see that the scale is going up in tens, using intervals of two. This arrow is therefore pointing to **134**

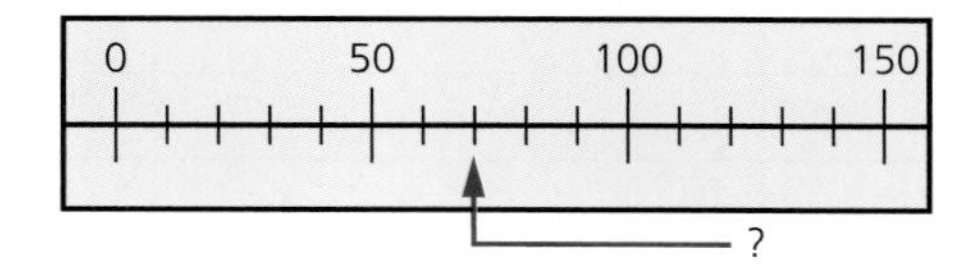

This one goes up in fifties, so each division is worth 10. The arrow here therefore points to **70**

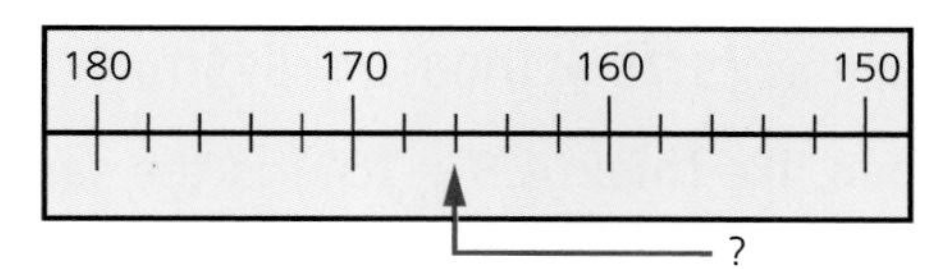

Attention! This one is going down not up! The scale is marked in tens and divided into twos, so the arrow here is pointing to **166**

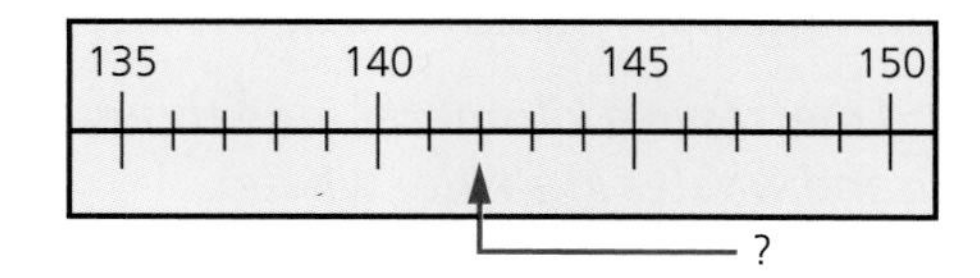

This scale is simply going up in fives, with each division representing one. The arrow here is therefore pointing to **142**

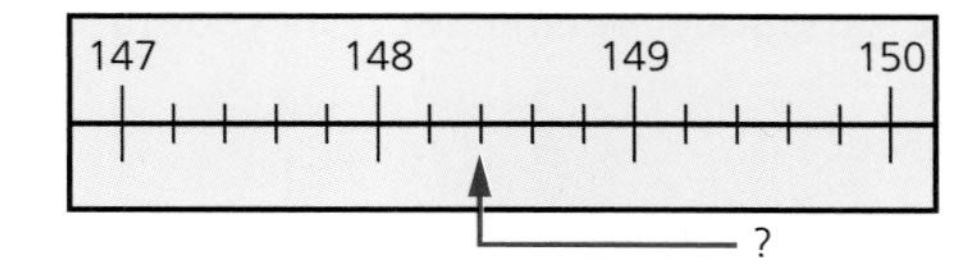

Finally a decimal scale: each unit is divided into 5 divisions, so each one represents 0.2 units (1 ÷ 5). The arrow here is pointing to **148.4**

The metric system

The metric system is based on powers of 10; that means 10, 100, 1000 and so on, as well as 0.1, 0.01, 0.001 etc. The scale goes outwards both ways.

Using a simple system of prefixes or 'multipliers', we can talk about any position on this scale. The actual unit of measurement is the same each time!

For	**length**	we use **metres** and the abbreviation	**m**
For	**mass**	**grams**	**g**
For	**time**	**seconds**	**s**
For	**capacity**	**litres**	**l**
For	**digital storage**	**bytes**	**B**

and there are a few others that scientists use. (Can you find any other examples?)

So what are the prefixes?

Effect	Put this in front	Say this
$\times 1$	(nothing)	(nothing)
$\times 1\,000$ ($\times 10^3$)	k	kilo
$\times 1\,000\,000$ ($\times 10^6$)	M	mega
$\times 1\,000\,000\,000$ ($\times 10^9$)	G	giga
$\times 1\,000\,000\,000\,000$ ($\times 10^{12}$)	T	tera
$\div 10$ ($\div 10^1$ or $\times 10^{-1}$)	d	deci
$\div 100$ ($\div 10^2$ or $\times 10^{-2}$)	c	centi
$\div 1\,000$ ($\div 10^3$ or $\times 10^{-3}$)	m	milli
$\div 1\,000\,000$ ($\div 10^6$ or $\times 10^{-6}$)	μ	micro
$\div 1\,000\,000\,000$ ($\div 10^9$ or $\times 10^{-9}$)	n	nano

Now let us try it out!

250 cm is the same as $(250 \div 100)$ m = 2.5 m

5.75 kg is the same as (5.75×1000) g = 5750 g

Write 4.8 cm in mm like this: $(4.8 \div 100)$ is the same as $(48 \div 1000)$ so it is 48 mm. Or just remember that 10 mm makes 1 cm and go from there.

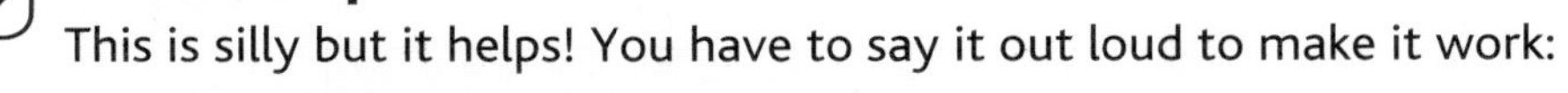

Revision tip

This is silly but it helps! You have to say it out loud to make it work:

'One thousand metres in a kilometre, one thousand grams in a kilogram and one thousand whales in a killer whale ...'

Common units

It is a very useful skill to be able to make reasonable estimates using units in common use. The guidance in this section should improve your reliability!

To estimate ...	think of ...
millimetre	window glass 3 mm thick
centimetre	width of little finger
inch	two fingers
foot	adult foot
yard	child's arm span
metre	doorway 2 m high
mile	15-minute walk
kilometre	3-minute bike ride
millilitre	medicine spoon, 5 ml
pint	pint of milk
litre	carton of orange juice

To estimate ...	think of ...
gram	milk bottle top, 2 g
ounce	egg, 2 oz
pound	three or four apples
kilogram	bag of sugar
stone	two house bricks
hundredweight	older child's weight
metric tonne	small car
ton	Range Rover, 3 tons
square cm	postage stamp, 5 sq cm
square yard	floor carpet, 20 sq yd
square metre	4 paving slabs

Metric and Imperial unit conversions

The system of measurement in use in Britain now is almost completely metric, but many examples of imperial measurements still exist (especially miles, feet, inches, pints, stone and tons), and you should be aware of them.

Metric	Imperial	
to convert	to	multiply by
millimetres	inches	0.0394
centimetres	inches	0.3937
centimetres	feet	0.0328
metres	inches	39.37
metres	feet	3.281
metres	yards	1.094
kilometres	yards	1094
kilometres	miles	0.621
grams	ounces	0.035
grams	pounds	0.0022
kilograms	pounds	2.205
kilograms	stone	0.158
kilograms	hundredweight	0.01968
kilograms	tons	0.00098
tonnes	tons	0.9842
square cm	square inches	0.155
square m	square yards	1.196
square m	acres	0.00025
hectares	acres	2.471
square km	square miles	0.386
cubic cm (ml)	fluid ounces	0.0352
litres	pints	1.76
litres	gallons	0.22
km per hour	miles per hour	0.625
m per second	miles per hour	2.2374

Imperial	Metric	
to convert	to	multiply by
inches	millimetres	25.4
inches	centimetres	2.54
feet	centimetres	30.48
inches	metres	0.0254
feet	metres	0.3048
yards	metres	0.914
yards	kilometres	0.0009
miles	kilometres	1.609
ounces	grams	28.35
pounds	grams	453.6
pounds	kilograms	0.4536
stone	kilograms	6.35
hundredweight	kilograms	50.8
tons	kilograms	1016
tons	tonnes	1.016
square inches	square cm	6.45
square yards	square m	0.8361
acres	square m	4047
acres	hectares	0.4047
square miles	square km	2.59
fluid ounces	cubic cm (ml)	28.41
pints	litres	0.568
gallons	litres	4.546
miles per hour	km per hour	1.6
miles per hour	m per second	0.4469

See the website http://www.aqua-calc.com for a full range of conversion tables and calculators.

You do not need to learn these conversions but it is useful to know a selection such as miles → km; lb → kg and inches → cm.

To convert degrees Fahrenheit to degrees Celsius: $C = \frac{5}{9}(F - 32)$

To convert degrees Celsius to degrees Fahrenheit: $F = \frac{9}{5}C + 32$

Perimeter, area and volume

Here is a reminder of the difference between these three words.

Perimeter – units of length, e.g. cm, m

Perimeter is the distance around the edge of a shape. Think of a fly walking all the way around it. How far would the fly walk?

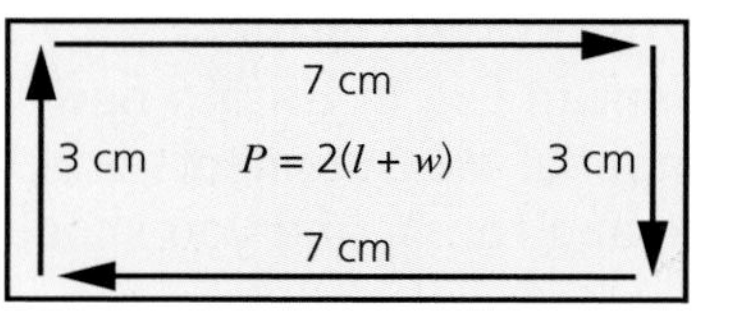

perimeter = 7 + 3 + 7 + 3 = 20 cm

Revision tip

'Perimeter' contains the word 'rim'!

Area

Area means counting squares inside the shape. If each square is 1 cm wide we say 'square cm'. We usually measure in units of square cm, square m or square km.

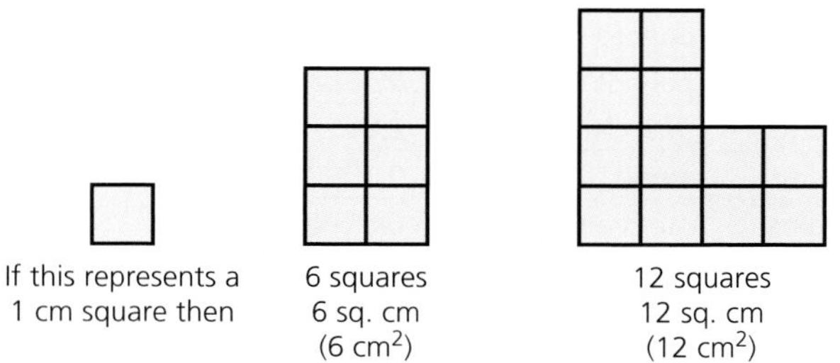

There is a formula that we can use to help us calculate the area of a rectangle.

Study these drawings of two rectangles.

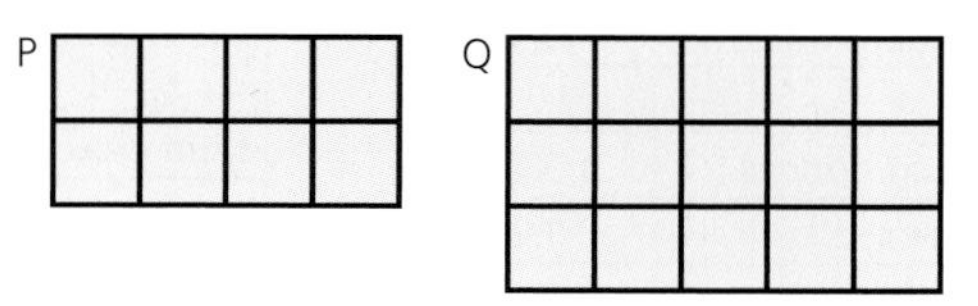

The area of rectangle P is 8 cm^2 (2 rows of 4 centimetre squares).

The area of rectangle Q is 15 cm^2 (3 rows of 5 centimetre squares).

You will notice that $2 \times 4 = 8$, and $3 \times 5 = 15$

We can write the formula for the area of a rectangle as $A = l \times w$ where A is the area, l is the length and w is the width of the rectangle.

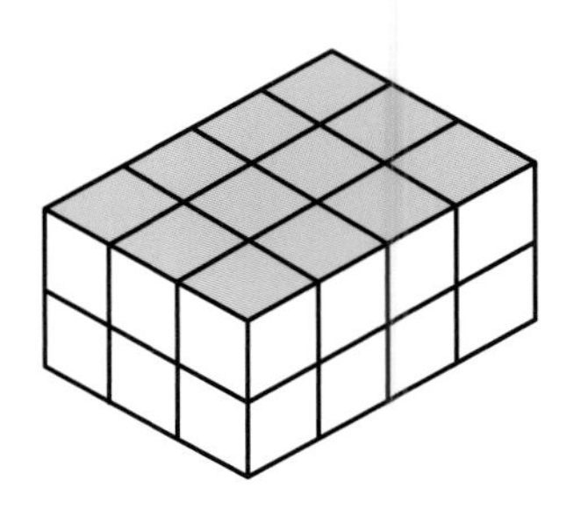

Volume

Volume can be thought of as **cubes** inside a solid shape. We often count in cubic centimetres (a cube 1 cm each way), but cubic metres are also useful.

Each cube in this **cuboid** box is one cubic centimetre.

It has two layers each of 12 cubes (4 × 3).

So the volume of the box is:

$2 \times 12 = 24$ cubic cm (24 cm^3)

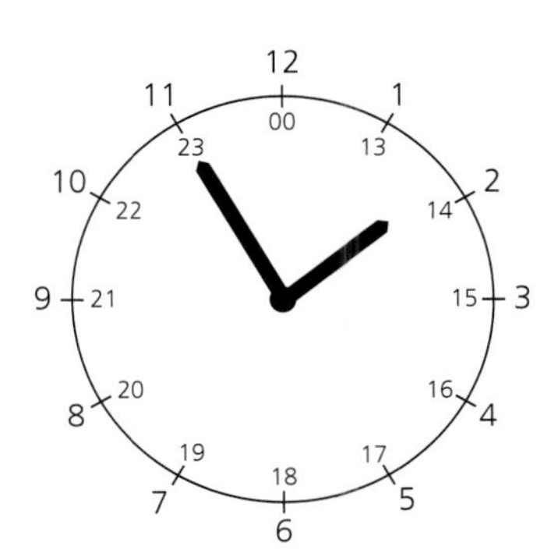

The 24-hour clock

Note that all you have to do is add 12 to the normal numbers on the clock to find the 24-hour clock equivalents.

Use a.m. (morning) and p.m. (afternoon) only when using the 12-hour clock.

When using the 24-hour clock always write the time using four digits.

The time shown on this clock depends on whether it is morning or afternoon.

morning	1.55 a.m. (12-hour clock)	01:55 (24-hour clock)
afternoon	1.55 p.m. (12-hour clock)	13:55 (24-hour clock)

Exam-style questions

Try these questions for yourself. The answers are given at the back of the book.

Some of the questions involve ideas met in earlier work that may not be covered by the notes in this chapter.

9.1 What are the readings on these scales?

(a) 10 20 30 40 (2)

(b) 50 60 70 80 90 100 110 (2)

9.2 This diagram shows two measuring beakers containing water.

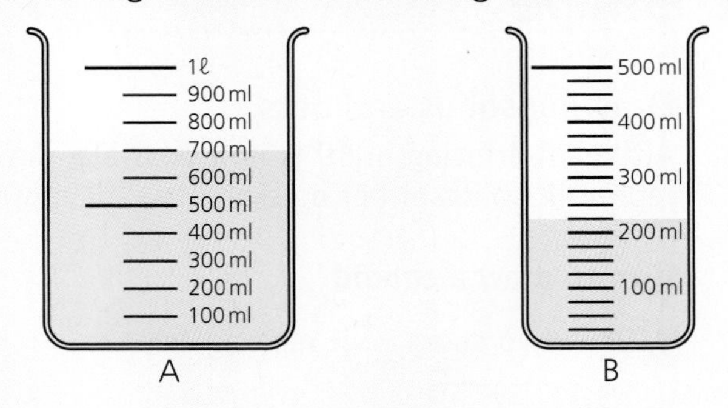

Write down, in millilitres, the volume of water in

(a) beaker A (2)

(b) beaker B. (2)

9.3 **(a)** The length of a desk is 123 cm.
Write this measurement in

(i) millimetres (1)

(ii) metres. (1)

(b) Tom's mass is 40 kilograms.

Write Tom's mass in

(i) grams (1)

(ii) pounds. (2)

9.4 A rectangle measures 6 cm by 5 cm. What is

(a) its perimeter (1)

(b) its area? (1)

9.5 The models shown below have been made of centimetre cubes.

A

B

What is the volume of

(a) model A (2)

(b) model B? (2)

9.6 (a) Write as 24-hour clock times:

(i) 1.10 p.m. (1)

(ii) 7.35 a.m. (1)

(b) Write as a.m./p.m. times:

(i) 14:30 (1)

(ii) 21:45 (1)

9.2 Shape

Drawing solids and nets

Although printing in 3D is now possible, drawing is still a 2D process! We represent solids either by sketching them or by constructing their nets.

How to draw a cuboid

- Draw two **congruent** rectangles.

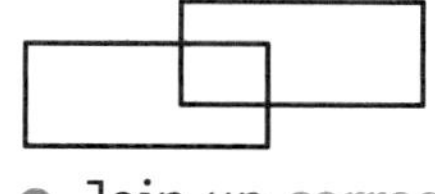

- Join up **corresponding** vertices.

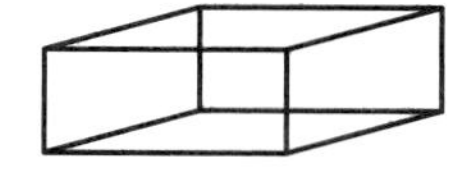

- Rub out the three hidden edges.

The most important rule is to make sure lines are parallel. In the second diagram there are three sets of parallel lines.

Nets

A net is a cardboard cut-out that can be folded up to make a solid shape. Here are three common examples:

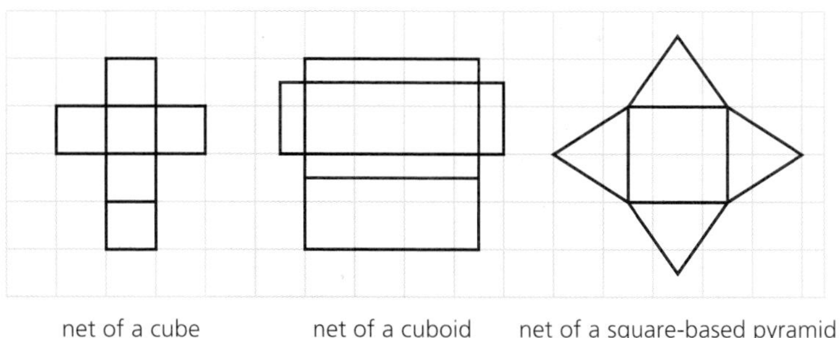

net of a cube net of a cuboid net of a square-based pyramid

Symmetry and congruence

What is the difference between **reflection** and **rotation** symmetry and what do we mean by the phrase *congruent shapes*?

Reflection (reflective) symmetry

This happy face has only one line of reflection symmetry (one mirror line).

You could place a mirror down the middle and see the hidden half of the picture in the reflection.

This would *not* work if you placed the mirror any other way.

Rotation (rotational) symmetry

These four strange birds are arranged in a pattern that has rotation symmetry. Here the symmetry is **four-fold** or **order 4** because there are four right ways up with this picture.

Another way of thinking of this is to see that the pattern fits exactly onto itself in four ways.

Patterns and shapes that have no rotational symmetry, such as the happy face on the previous page, are described as having rotational symmetry of order 1 (not zero). This is because it has exactly one right way up, or only one way to fit onto itself.

This triangle has *no* mirror lines, but it does have three-fold rotation symmetry. This is because the blue and white pattern in the triangle would break up any possible reflection symmetry.

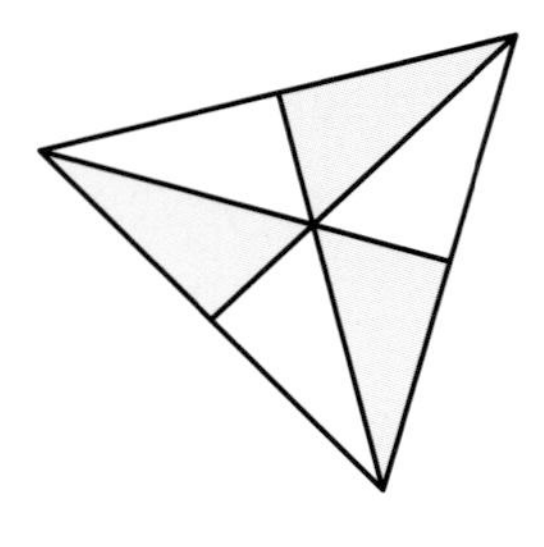

Congruence

Two shapes are described as **congruent** if one is an exact copy of the other. It does not matter if one is rotated or reflected, provided the size and shape of both are identical. (If only the sizes are different, we say that the shapes are **similar**.)

congruent shapes

Types of triangle, types of quadrilateral

Triangles and **quadrilaterals** are so common that we sort them into different categories according to their properties.

Triangles

- **Equilateral** triangles have all the angles the same (each 60°) and all sides equal length.
- **Isosceles** triangles have two equal angles and two equal sides.
- **Scalene** triangles have no equal angles and no equal sides.
- **Right-angled** triangles have one angle of 90°.

  A right-angled triangle may be scalene or isosceles.

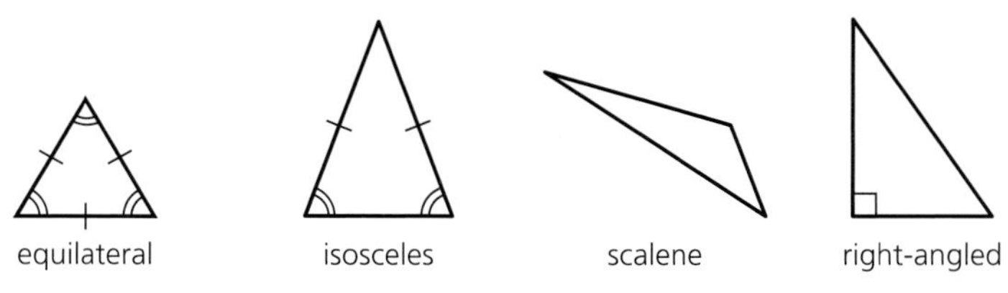

Revision tip

To help you remember what an isosceles triangle is, sing this verse to the tune 'O Christmas Tree': 'Isosceles, Isosceles, Two angles have equal degrees' – to help you remember how to spell the word, say 'I Sat On Seagull Cliff Eating Lumpy Egg Sandwiches'.

Quadrilaterals

- A **square** has four equal sides and four equal angles (each 90°). Opposite sides are parallel.
- A **rectangle** (oblong) also has four equal (right) angles, but only opposite sides are equal length. Opposite sides are parallel.
- A **rhombus** (diamond) has four equal sides, but only opposite angles are equal. Opposite sides are parallel.
- A **parallelogram** has two pairs of parallel sides. Opposite sides are not only parallel but also equal in length. Opposite angles are equal.
- A **kite** has two pairs of adjacent equal sides and two equal angles (including delta).
- A **trapezium** has one pair of parallel sides. In an isosceles trapezium the non-parallel sides are equal in length.

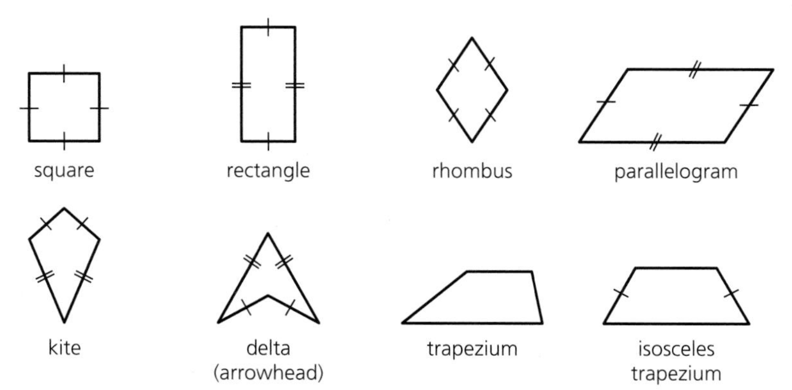

Exam-style questions

Try these questions for yourself. The answers are given near the back of the book.

Some of the questions involve ideas met in earlier work that may not be covered by the notes in this chapter.

9.7 The cuboid represented by this drawing has been made from cardboard marked with centimetre squares.

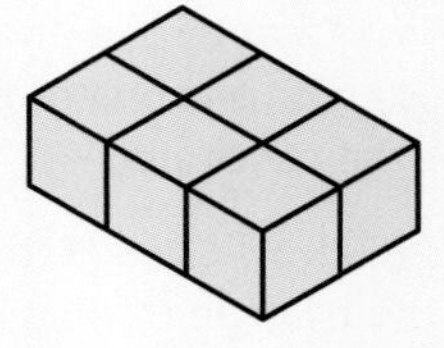

(a) Draw an accurate net for the cuboid. (3)

(b) What area of cardboard is needed to make the cuboid? (Edges are glued together without any 'tabs'.) (3)

9.8 Two quadrilaterals are drawn here.

(a) Name the shapes. (2)

(b) Draw all lines of symmetry on the shapes. (3)

(c) Describe the rotational symmetry of the shapes. (2)

9.9 Look carefully at the shapes drawn on this grid.

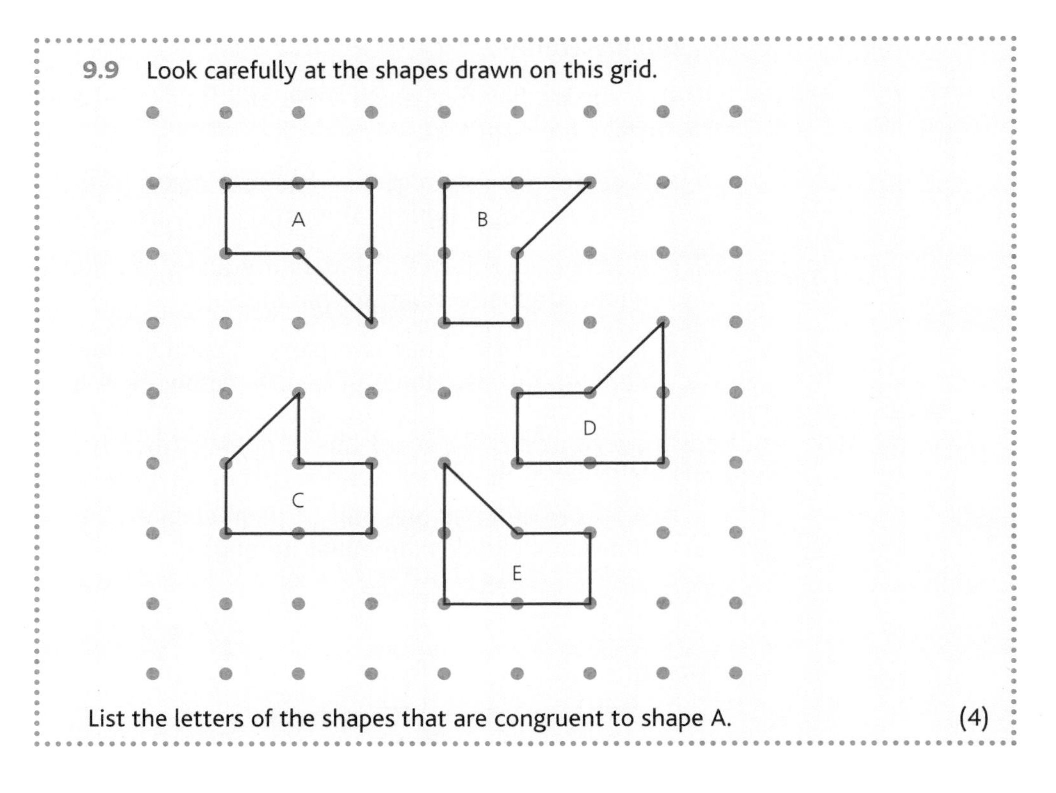

List the letters of the shapes that are congruent to shape A. (4)

9.3 Space

Triangle construction

When you **construct** (draw) accurate scale diagrams you should be accurate to within one millimetre and one degree. This is how to construct triangles using just the information given.

Three sides

Example

Draw a triangle of side lengths 3 cm, 5 cm and 7 cm.

First draw the longest side at the bottom of the page.

Place the point of a pair of compasses at one end of this line and draw an **arc** whose radius is one of the other two lengths. Repeat for the other side.

The point where the two arcs (curves) cross is the **apex** (top) of the triangle. Join up the triangle using a ruler.

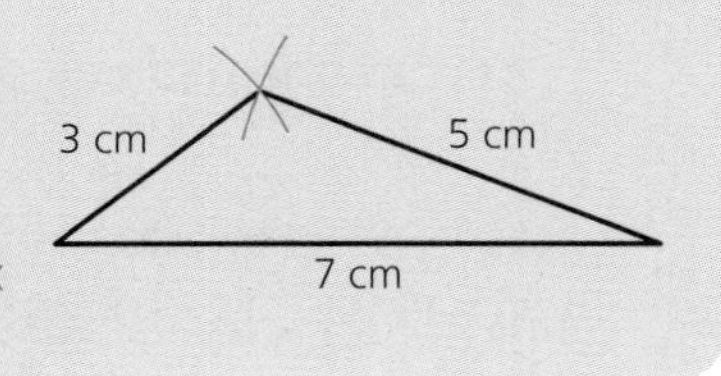

Two sides and the angle between them

Example

Draw a triangle in which two of the sides measure 7 cm and 5 cm. The angle between them is 35°.

Draw the longer of the two sides horizontally on the page and measure out the given angle at one end. Extend this second side a little further than you think you will need. Now measure along this second side for the correct distance before closing up the triangle.

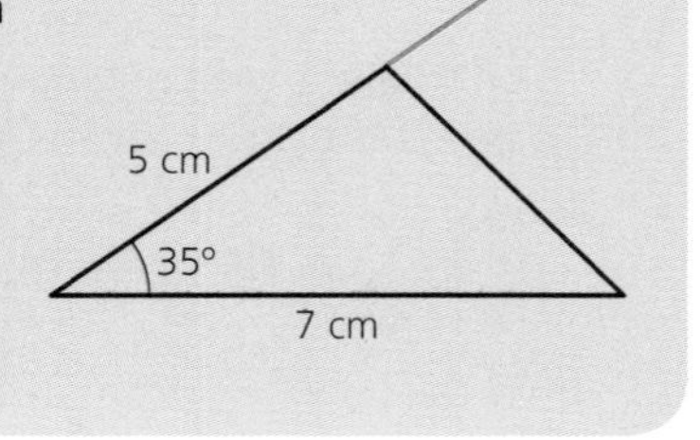

One side and the angle at each end

Example

Draw a triangle with base 4 cm and base angles 60° and 30°.

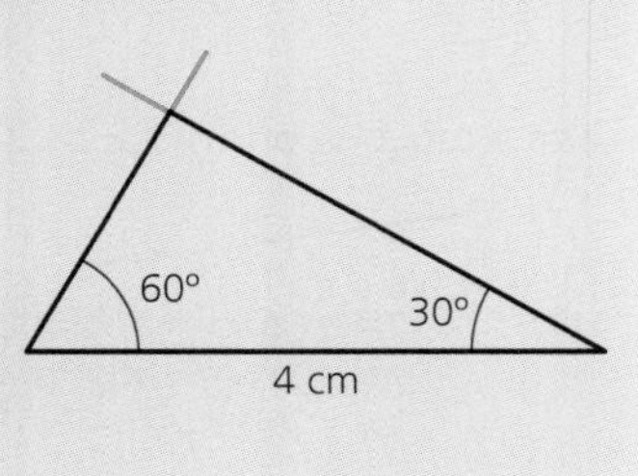

This is rather like the situation above, but with two angles drawn rather than one. Here, however, only one length has to be measured.

Draw the given side horizontally on the page and construct the desired angle at each end. Continue these two new lines until they cross – the **intersection** point is the apex of the triangle.

Compass constructions

A ruler (straight edge) and a pair of **compasses** are all you need to make simple **constructions**.

Revision tip

Your skills with handling compasses will improve dramatically if you take time to draw accurate circle pictures such as this one.

Bisecting an angle

Bisecting means cutting into two equal parts. It is possible to **bisect** an angle without any measuring. This is the procedure:

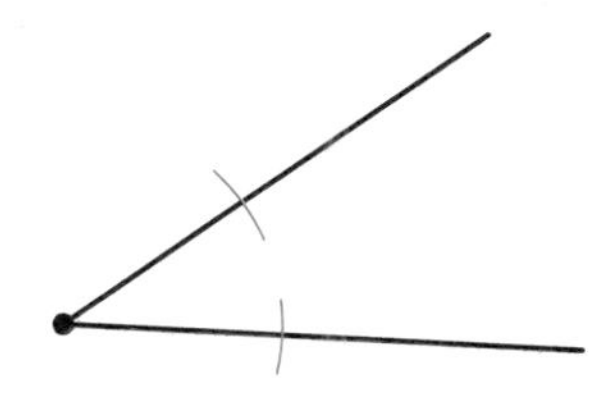

1 With the compass point at the vertex (hinge) of the angle, make an arc on each arm (side) of the angle.

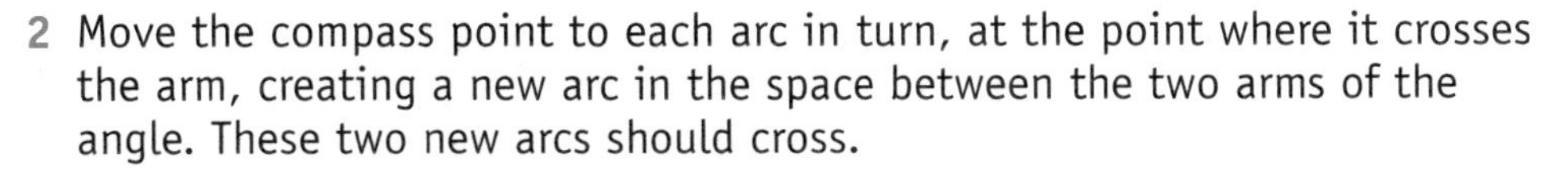

2 Move the compass point to each arc in turn, at the point where it crosses the arm, creating a new arc in the space between the two arms of the angle. These two new arcs should cross.

3 Join the crossing point of the two new arcs to the vertex of the angle with a ruler to complete the bisection.

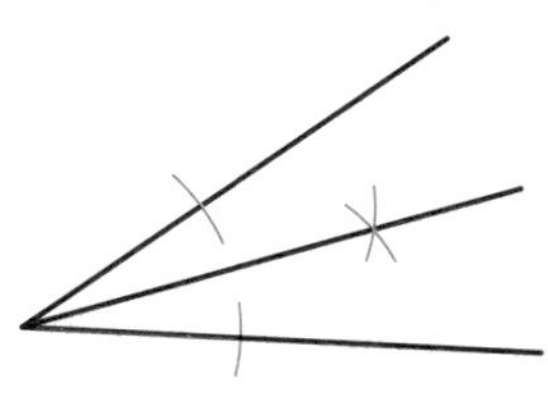

Bisecting a line segment

A line, in theory, goes on for ever. A line segment is therefore a more accurate way to describe a line with two ends. Even if we do not know the length of this line segment, we can still bisect it (cut it in half exactly) using just compasses and a straight edge (ruler without markings). This is the procedure:

1 Without adjusting the compasses, place the point at each end of the line segment and make an arc above and below it. The arcs made from one end should cross the arcs made from the other.

2 Using a ruler, join the points where the arcs crossed above the line segment to the point where they crossed below it. This straight line will go through the centre of the line segment.

Note that the two line segments intersect at right angles, so this is also called a perpendicular bisector.

Types of angle

Angles can be any size from 0° to 360°. They are given special names according to their size.

Size of angle (degrees)	Name
Less than 90	Acute
90	Right
90 to 180	Obtuse
180	Straight
180 to 360	Reflex

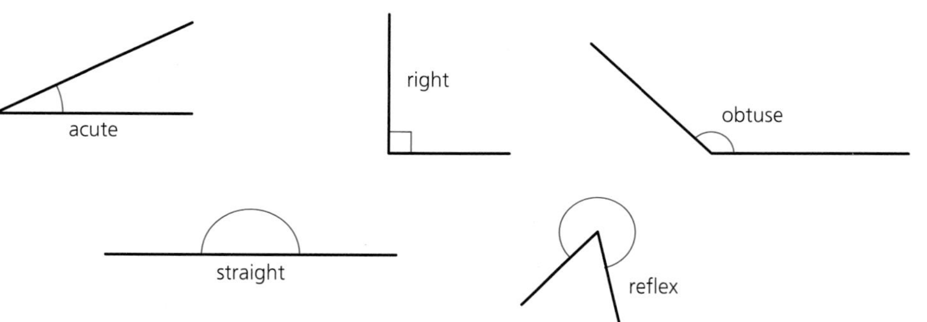

Revision tip

Think of a mnemonic for remembering the order of these names: e.g. Acute, Obtuse, Reflex are in alphabetical order, or 'Always Run Over Stupid Robots!'

Knowing the approximate size of each type of angle helps with estimation, especially in bearings questions. It is good practice to sketch out a scale diagram in rough before you draw it accurately with the exact angles, to make sure you don't run out of paper!

The eight-point compass

We use a compass to help us describe the direction of a journey.

Revision tip

How do you remember the points of the compass? People have suggested, working clockwise, 'Naughty Elephants Squirt Water' or even 'Nice Easy School Work!' Perhaps the best is simply to remember 'North East South West'.

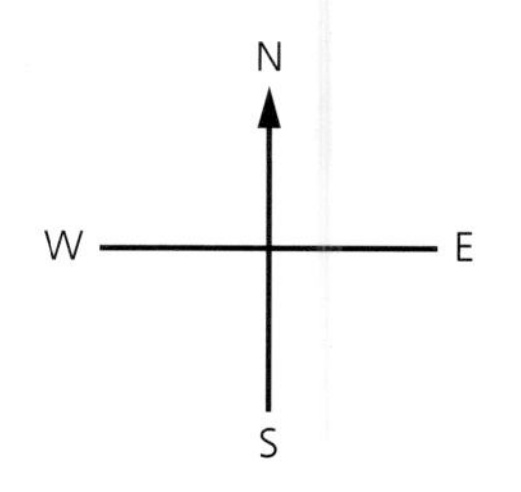

We can then add the diagonals to this to give us four more directions.

Note that the diagonal directions use the two letters either side of them, but always starting with N or S.

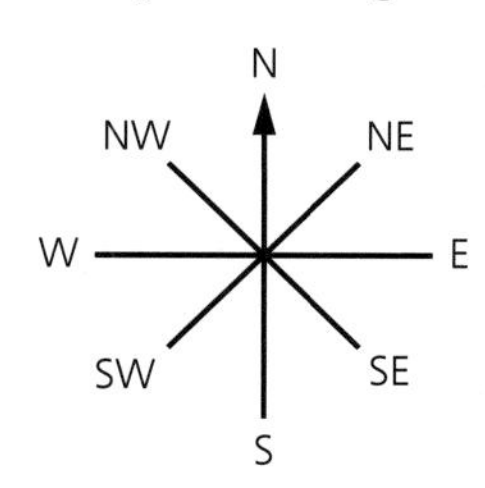

To get from Newhaven to Dieppe, the boat has to sail in a south-east direction for 120 km.

On the way back it sails in a north-west direction for 120 km.

(120 km is about 75 miles.)

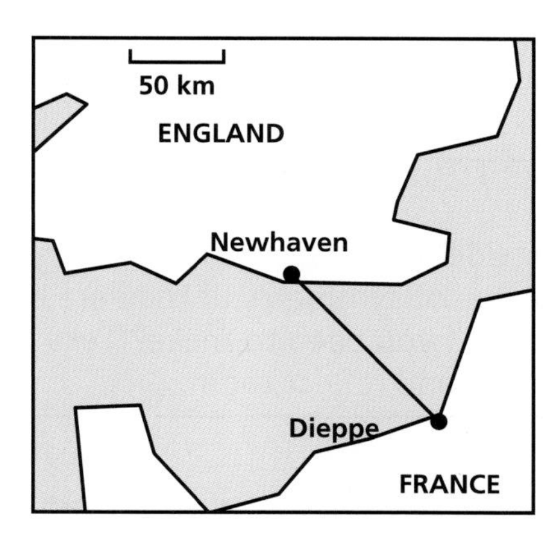

Compass points

Bearings are always given clockwise from north, and are always given as a three-digit number.

Try to get used to estimating angles before drawing or measuring them, in order to prevent the mistake of reading the wrong scale on your angle measurer or protractor.

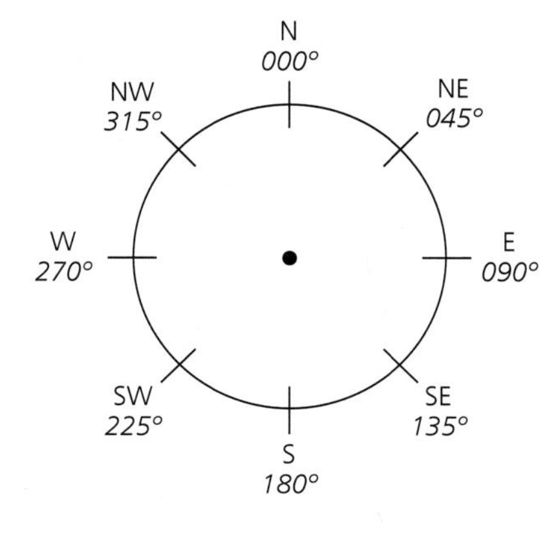

Reflections on diagonal lines

Our eyes are good at looking at symmetry that is lateral (left-right) but we find it much harder if it goes up or down (vertical). Symmetry on a diagonal is even harder.

To make it easier, just turn the paper through 45° and the symmetry line is now vertical! If the symmetry line is now horizontal then you turned it the wrong way. Go back and turn it in the other direction!

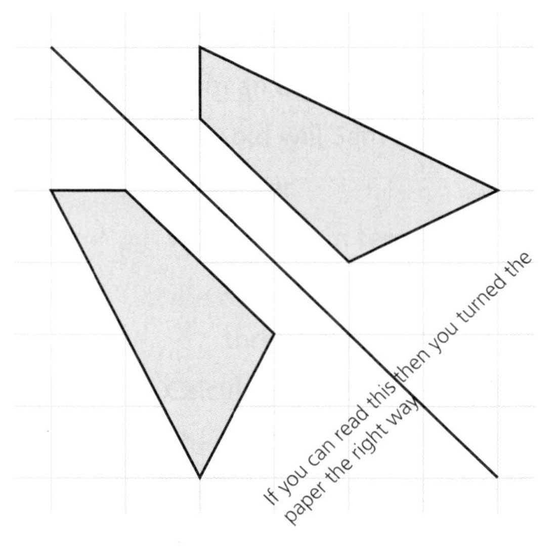

Revision tip

Look around outside for things with symmetry, especially things that have a diagonal line of symmetry. If they are not quite symmetrical, what changes would you need to make? Take photographs and use a computer to rotate and reflect them.

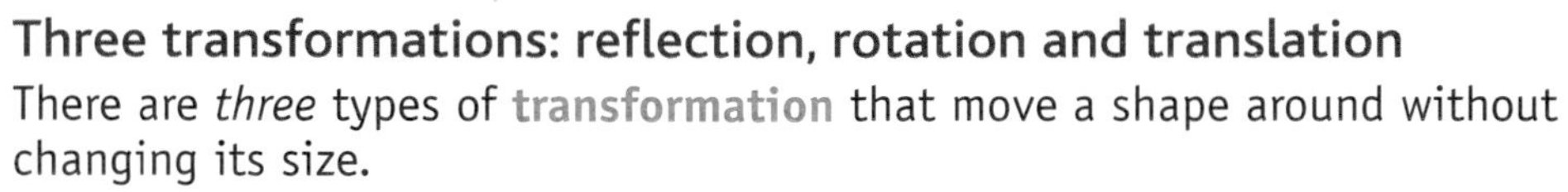

Three transformations: reflection, rotation and translation

There are *three* types of **transformation** that move a shape around without changing its size.

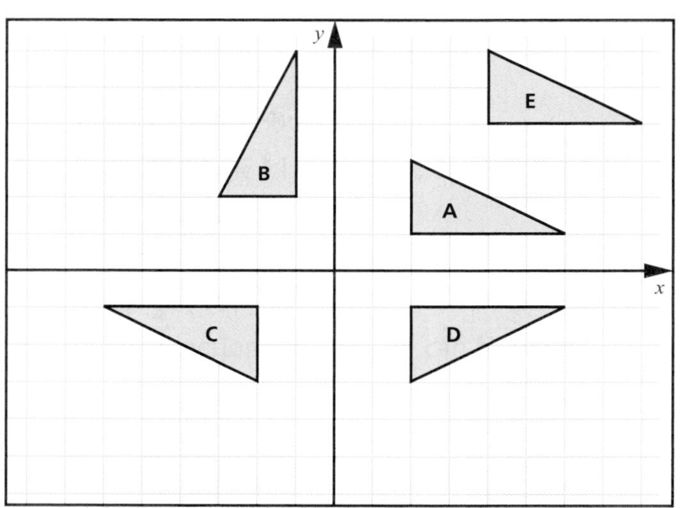

The three transformations used above were reflection, rotation and **translation**:

To map **A** on to **B** the transformation is rotation 90° anti-clockwise about (0, 0).

To map **A** on to **C** the transformation is rotation 180° about the origin.

To map **A** on to **D** the transformation is reflection in the x-axis.

To map **A** on to **E** the transformation is translation two units right and three units up.

Exam-style questions

Try these questions for yourself. The answers are given near the back of the book.

Some of the questions involve ideas met in earlier work that may not be covered by the notes in this chapter.

9.10 **(a)** Construct triangle ABC where AB is 6 cm, BC is 5 cm and AC is 4 cm. (3)

(b) Measure, to the nearest degree, angle BAC. (1)

9.11 Triangle PQR is drawn here.

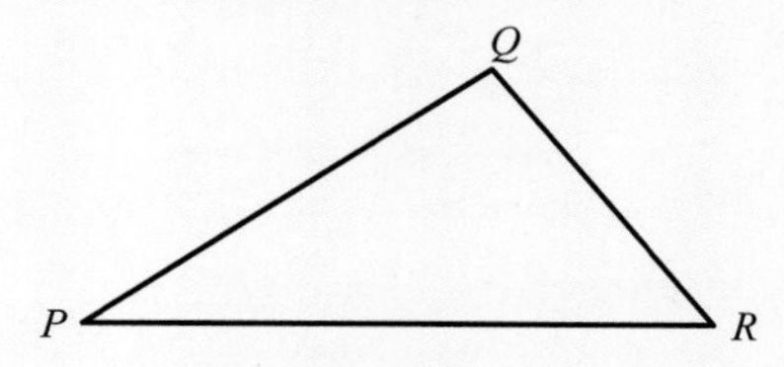

(a) What type of angle is

(i) angle PQR (1)

(ii) angle PRQ? (1)

(b) What name is given to this type of triangle? (1)

9.12 The diagram below shows four pupils standing on a patio.

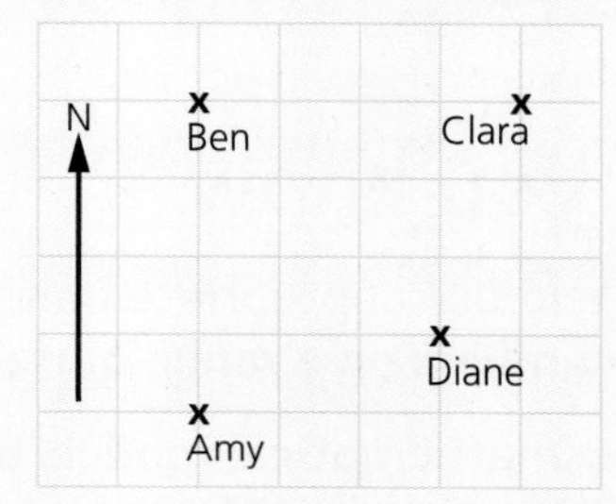

Using the eight points of the compass, describe the direction of

(a) Ben from Amy (1)

(b) Amy from Ben (1)

(c) Clara from Amy (1)

(d) Diane from Ben. (1)

9.13 Reflect the shape in the line. (3)

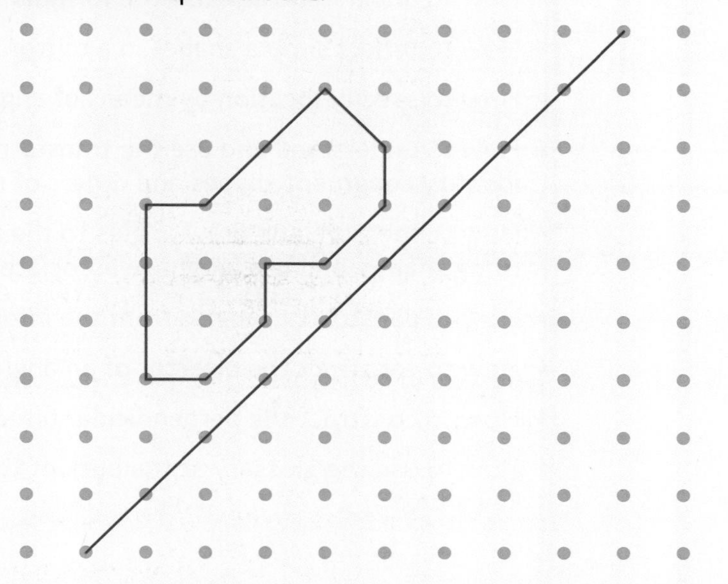

9.14 On the grid below:

(a) rotate shape A through 90° clockwise about the black dot (2)

(b) translate the shape A two units right and three units down. (2)

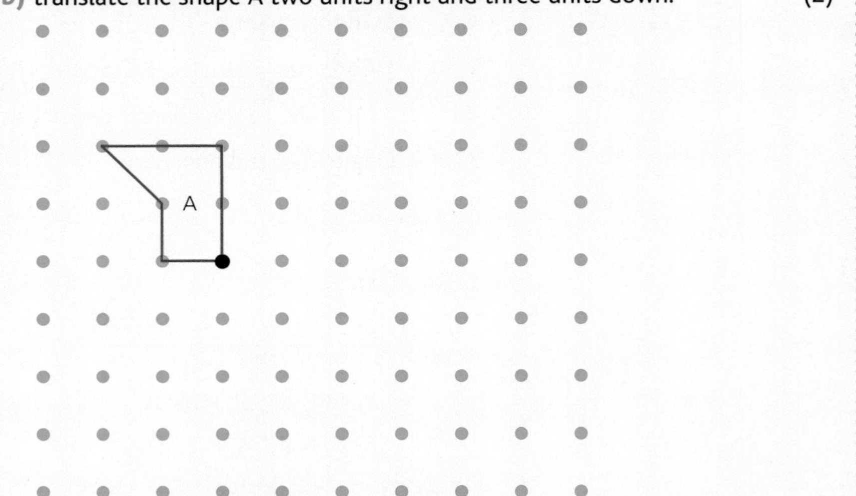

★ Make sure you know

- ★ How to choose and use appropriate units and instruments, interpreting, with accuracy, numbers on a range of measuring instruments
- ★ How to make 3D mathematical models by linking given faces or edges; draw common 2D shapes in different orientations on grids
- ★ How to find perimeters of simple shapes and find areas by counting squares
- ★ How to find volumes by counting cubes
- ★ The rough metric equivalents of imperial units still in daily use and how to convert one metric unit to another
- ★ How to understand and use the formula for the area of a rectangle
- ★ How to reflect simple shapes in a mirror line
- ★ How to specify location by means of angle and distance
- ★ How to understand and use the transformations rotation and translation; identify congruent shapes and orders of rotational symmetry
- ★ How to measure and draw angles to the nearest degree, when drawing or using shapes, and use language associated with angles
- ★ How to construct triangles from the information given
- ★ How to construct the bisector of an angle
- ★ How to construct the perpendicular bisector of a line segment
- ★ How to use the glossary at the back of the book for definitions of key words

Test yourself

Before moving on to the next chapter, make sure you can answer the following questions. The answers are at the back of the book.

1 What are the readings on these scales?

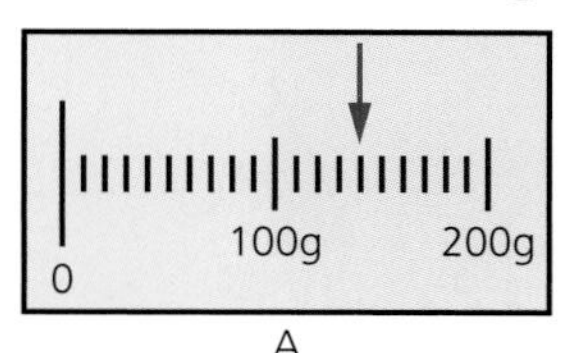

A

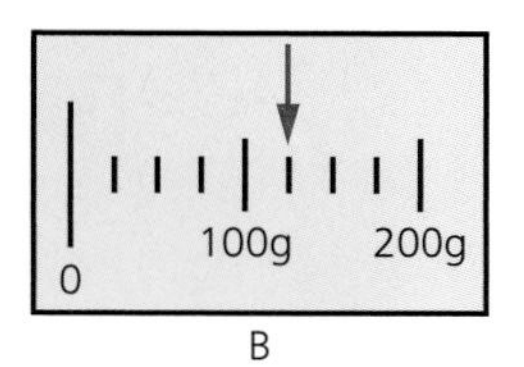

B

2 The shape below has a perimeter of 14 cm and an area of 6 cm².

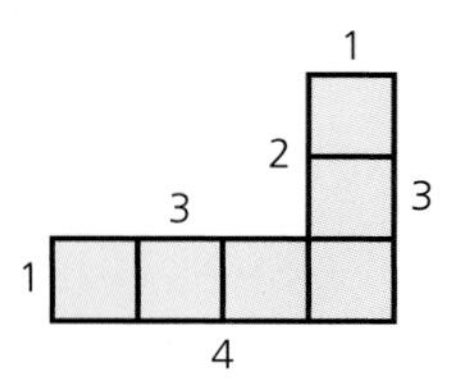

(a) Draw a shape which has perimeter 10 cm and area 6 cm².

(b) Draw a shape which has perimeter 14 cm and area 10 cm².

3 This model has been made from interlocking centimetre cubes.

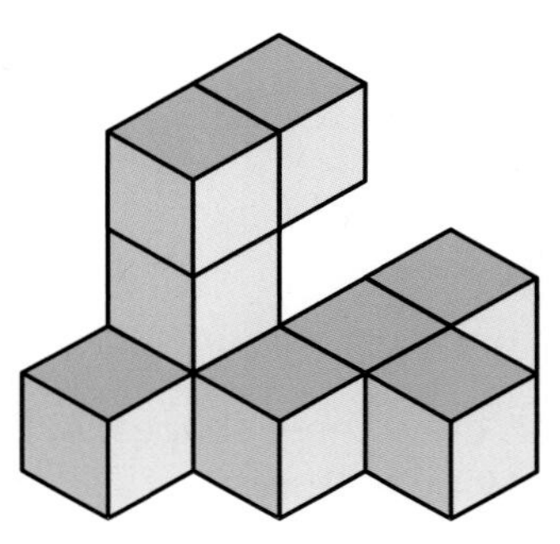

(a) What is its volume?

(b) On isometric paper, draw a different view of the model.

4 **(a)** Construct triangle ABC where AB is 10 cm, angle BAC is 30° and angle ABC is 40°.

(b) Measure and write down

(i) angle ACB

(ii) AC

5 Five congruent shapes are drawn on this grid.

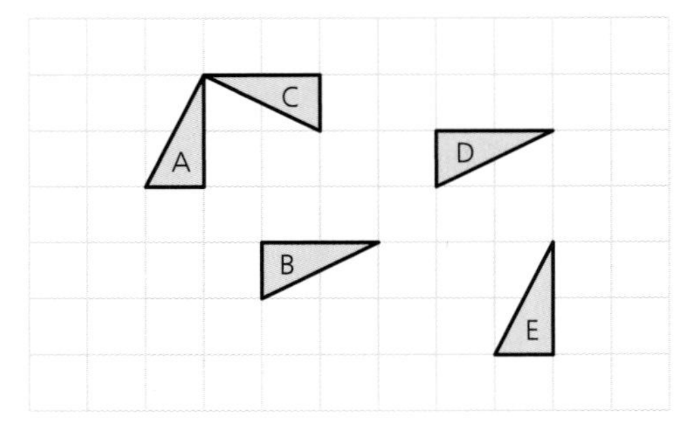

(a) What does **congruent** mean?

(b) Which shape could be reached from shape A by:

(i) reflection

(ii) rotation

(iii) translation?

6 The square below has area 16 cm², 4 lines of symmetry and rotational symmetry of order 4

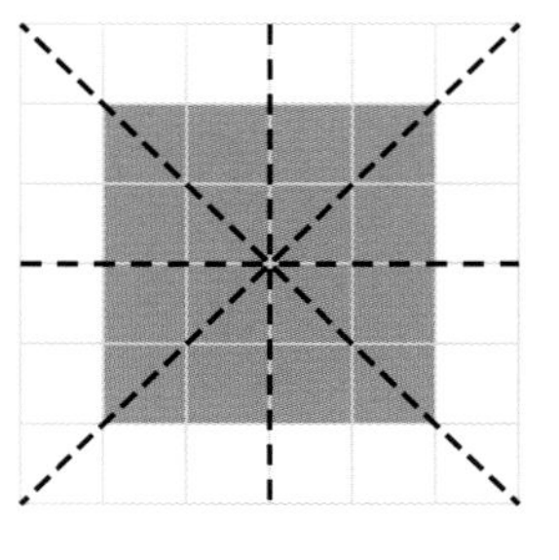

On squared paper, draw three shapes, A, B and C, all with area 16 cm², and with the following features:

Shape A one line of symmetry only

Shape B two lines of symmetry

Shape C rotational symmetry of order 4 but no lines of symmetry

Hint: start with a square each time and modify it.

Geometry and measures (2)

10.1 Measures

In this section we review circles, area, volume and speed.

Parts of a circle and the value of pi

The **circumference** of the circle is the same as its perimeter – it is the distance all the way around the edge.

The **diameter** is any straight line that goes from one side of the circle to the other, through the centre. The diameter divides the circle into two equal halves.

The **radius** is any straight line that runs from the centre of the circle to the circumference.

A **chord** is a straight line joining any two points on the circumference.

A **segment** is part of a circle cut off by a chord.

A **sector** is part of a circle cut off by two radii.

An **arc** is a section of the circumference.

A **tangent** is a straight line that just touches the circumference.

A **semicircle** is half a circle, cut at the diameter.

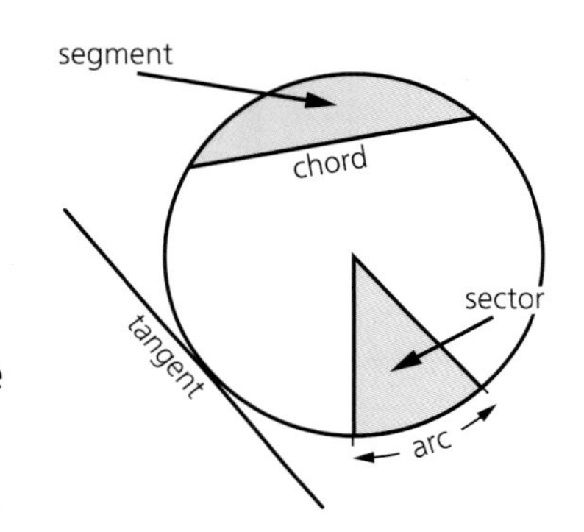

Pi

Pi (π) is the famous number that is calculated by dividing the circumference of any circle by its diameter. Its approximate value is 3.14 or $\frac{22}{7}$ or $\frac{355}{113}$ or even $\frac{312\,689}{99\,532}$!

Let's give it a few more decimal places. How about the first fifty?

π = 3.141 592 653 589 793 238 462 643 383 279 502 884 197 169 399 375 10 ... (How many can you memorise, for fun?)

Areas of circles

Start by reminding yourself of the words used to describe the parts of a circle in the earlier diagrams.

Abbreviation	Meaning
r	Radius
d	Diameter
C	Circumference
A	Area
h	Height (of a cylinder for example)

(Note that we use capital letters for C and A)

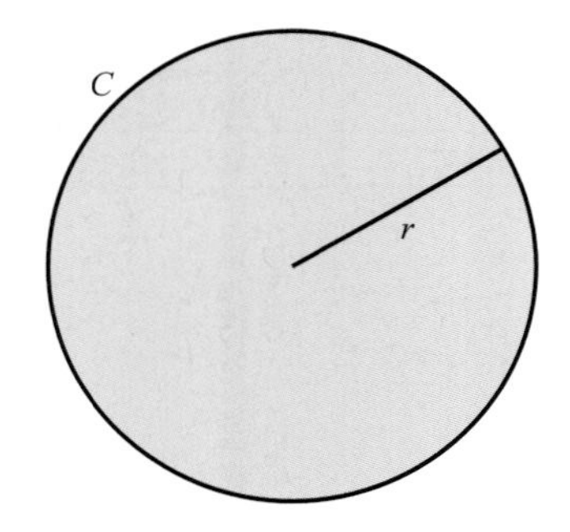

You need to learn these formulae:

$d = 2r$ $C = \pi d$ $C = 2\pi r$ $A = \pi r^2$

volume of cylinder $V = \pi r^2 h$

If there is no π (pi) button on your calculator, then you can use 3.14 instead.

Level 3
questions may involve finding the radius of a circle of given circumference or area.

Example

How far along the ground will a wheel of radius 30 cm travel if it turns 1000 times?

The radius is 30 cm (0.3 m).

So the circumference (one turn of the wheel) is $2 \times \pi \times 0.3$ which comes to 1.884 96 m.

Thus after 1000 turns, the wheel has travelled 1884.96 m or about 1.9 km.

Level 1
candidates will be given a sheet of formulae for Common Entrance.

Level 2 and 3
candidates need to learn the formulae.

Area and volume formulae

Area formulae

Square

$A = a^2$

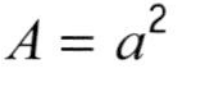

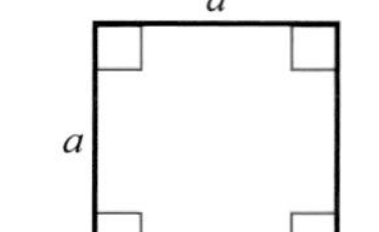

Rectangle

$A = lw$

Triangle

$A = \frac{1}{2}bh$

Parallelogram

$A = bh$

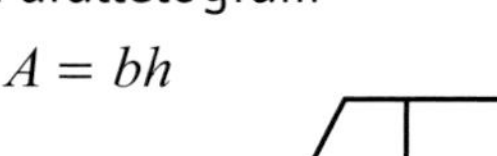

Trapezium

$A = \frac{1}{2}(a + b)h$

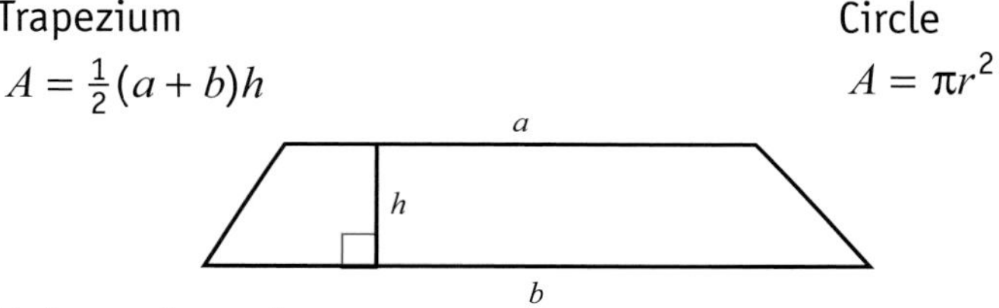

Circle

$A = \pi r^2$

Volume formulae

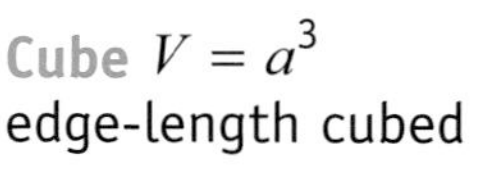

Cube $V = a^3$
edge-length cubed

Cuboid $V = lwh$
length × width × height

Prism volume = cross section area × height (length).
This includes cylinders!

Pyramid volume = $\frac{1}{3}$ × base area × height.
This includes cones!

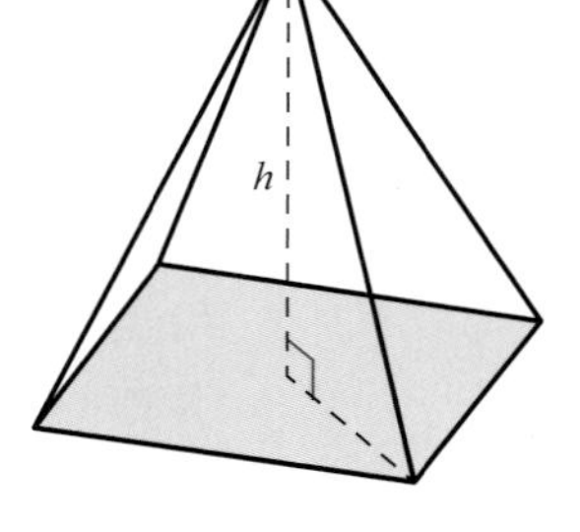

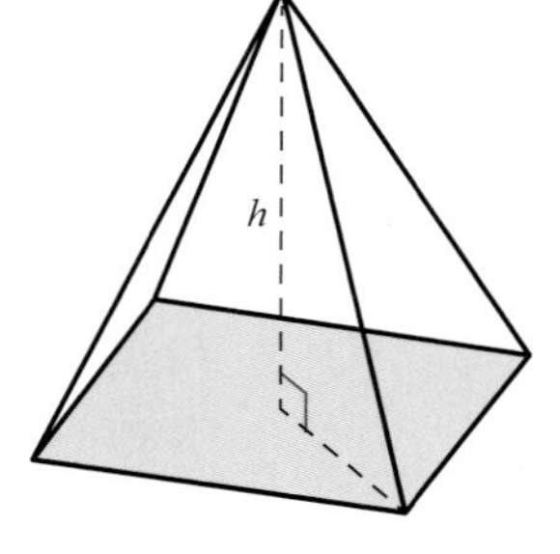

Level 1
questions may involve calculating:
- the areas of triangles and parallelograms
- the volumes of cuboids.

Calculating areas of fractional and composite shapes

Most shapes in everyday life are not simple but all are made up of simple shapes in some way.

Fractional shapes

These are ordinary shapes such as rectangles and circles, but with holes cut out or pieces missing. You calculate the area of a fractional shape using subtraction.

The path around this small garden is 1 m wide. The area of the path is found by subtracting the area of the garden from the area of the surrounding rectangle.

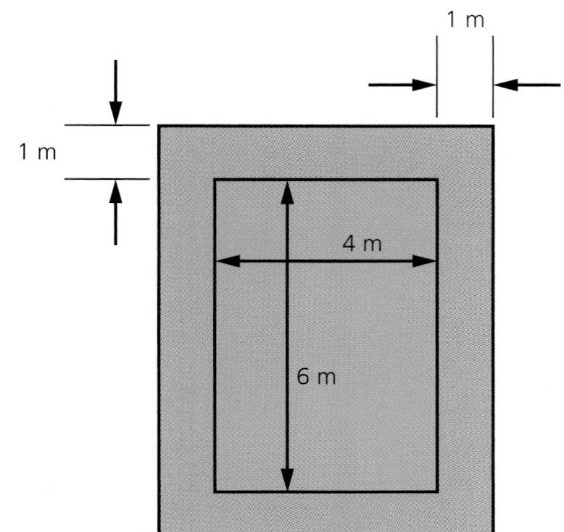

area of garden = 4 × 6

= 24 m^2

area of surrounding rectangle =
(1 + 4 + 1) × (1 + 6 + 1) = 48 m^2

so area of path = 48 – 24

= 24 m^2

Composite shapes

These shapes are made by joining together simpler shapes such as squares and triangles. You calculate the area of a composite shape using addition.

With composite shapes, it is usually up to you to work out the best way of dividing the area up into simpler shapes.

Of course, it doesn't actually matter which way you do it – they all give the same answer!

This archway looks complicated but it is really just made up of simpler shapes, as the second diagram clearly shows.

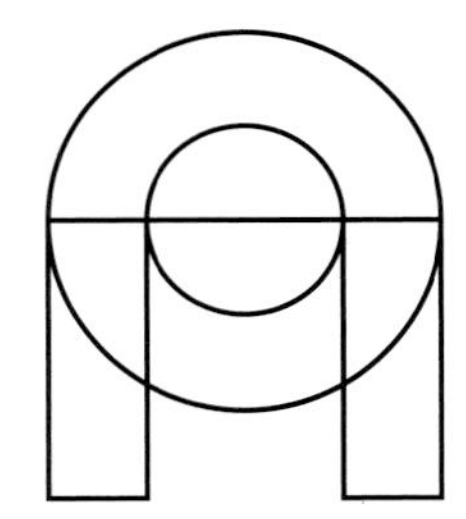

When working out its area, we would subtract the area of the small circle from the area of the large circle and then halve the answer to find the area of the curved part of the arch. To this we would add the area of the two rectangular feet and obtain the final answer.

Calculating the volumes of fractional and composite shapes

Check that you are familiar with the standard volume formulae above, especially of the prism. Here is one more:

sphere volume = $\frac{4}{3}\pi r^3$

Example 1

A cake in the shape of a cylinder (height 7 cm, radius 12 cm) is cut into sectors of 60 degrees each. Find the volume of each piece.

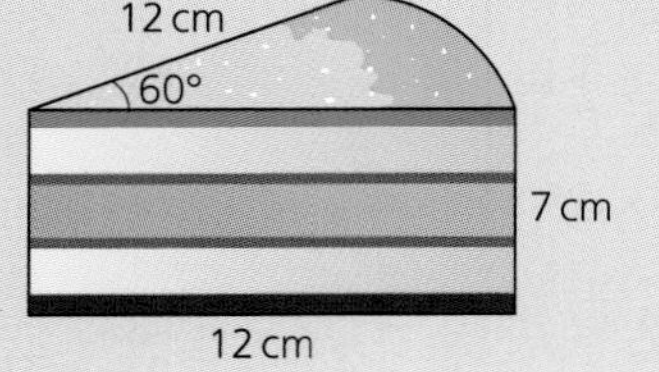

$360° \div 60° = 6$, so each piece is $\frac{1}{6}$ of the cake

volume of whole cake (cylinder) $= \pi r^2 h$

so volume of cake is $\pi \times 12^2 \times 7 = 3166.725$ cm^3

thus volume of piece $\frac{3166.725}{6} = 527.8$ cm^3

Example 2

The base of this toy is a hemisphere (half a sphere).

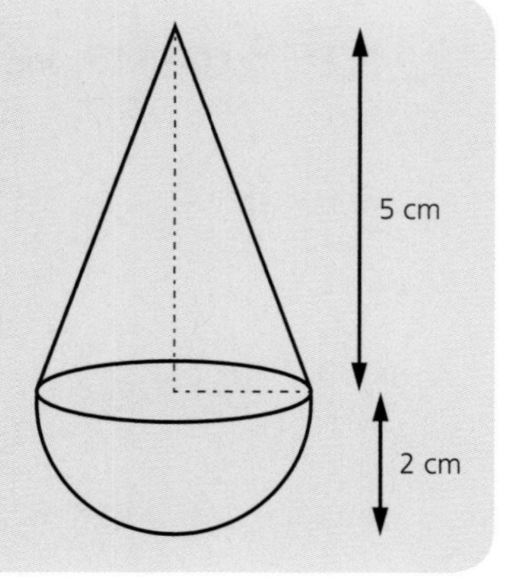

To find the volume of the toy we add up the volumes of the upper cone and the lower hemisphere:

volume of hemisphere $= \frac{1}{2} \times \frac{4}{3} \times \pi \times 2^3$

$= 16.76$ cm^3

volume of cone $= \frac{1}{3} \times (\pi \times 2^2) \times 5$

$= 20.94$ cm^3

so volume of toy $= 37.7$ cm^3

Distance, speed and time

Distance, speed and time are closely related, and so questions will usually ask you to find any one of the three.

If you learn the 'magic triangle' below, then you will surely become a magician at working out such problems!

Simply put your finger over the one you want to find and the magic triangle tells you what calculation to do with the other two. For example, to find speed you have to do distance ÷ time.

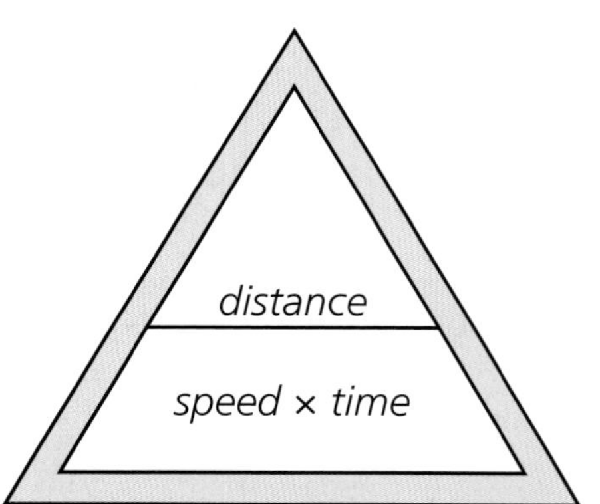

Example

How long does it take to travel 80 miles if you are travelling at 50 miles per hour?

time = distance ÷ speed

= 80 ÷ 50

= 1.6 hours

(OK, but what does the 0.6 hours mean?)

0.6 hours = 0.6 × 60 minutes

= 36 minutes

So the journey takes 1 hour 36 minutes.

Speed is an example of a **compound measure.** Other compound measures, such as **density** and **pressure**, may be met in your science studies.

Note: Mathematics papers at all levels will test only speed.

Level 3
Level 3 questions may involve average speeds of multi-stage journeys.

Exam-style questions

Try these questions for yourself. The answers are given at the back of the book. For questions requiring the use of a calculator, give your answers to 3 s.f. unless otherwise directed.

Some of the questions involve ideas met in earlier work that may not be covered by the notes in this chapter.

10.1 (a) The length of Tom's pencil is 13 cm.

(i) Write this length in millimetres. (1)

(ii) Write this length in metres. (1)

(iii) What is this length, approximately, in inches? (1)

(b) Mr Smith is 6 feet tall.

What is his approximate height in metres? (1)

(c) 5 miles is approximately the same as 8 kilometres.

The road distance between John O'Groats and Land's End is about 600 miles.

What is this distance in kilometres? (2)

10.2 (a) A field has an area of 50 000 square metres.

What is this in hectares (ha)? (1)

(b) A cube has edges of length 10 cm.

(i) What is the area of one face in square centimetres (cm^2)? (1)

(ii) What is the volume of the cube in cubic centimetres (cm^3)? (1)

10.3 (a) A beetle crawled 36 centimetres in 3 minutes.

(i) What was its average speed in centimetres per minute? (1)

(ii) What was the average speed in cm/s? (1)

(iii) How far, at this average speed, could the beetle crawl in an hour? (1)

(b) Mr Williams drives at an average speed of 63 mph (miles/h) on a motorway.

(i) How many miles does he travel in 10 minutes? (2)

He passes the Yummy Snack service station at 09:00

(ii) At what time will he pass the Last Chance service station which is 21 miles away from the Yummy Snack? (2)

10.4 (a) (i) Construct triangle ABC with $AB = 9$ cm, $BC = 7.5$ cm and $AC = 8$ cm. (4)

(ii) Measure and write down the shortest distance from C to AB. (2)

(b) A square has diagonals of length 10 centimetres.

(i) Make an accurate drawing of the square. (3)

(ii) By taking suitable measurements, find the perimeter of the square. (2)

10.5 **(a)** On an isometric grid, draw a representation of a cuboid measuring 7 cm by 5 cm by 4 cm. (2)

(b) Calculate the volume of the cuboid. (2)

(c) Calculate the total surface area of the cuboid. (4)

10.6 Felix is making a scale model of a battleship, using a scale of 1 : 250

(a) If the length of the battleship is 90 metres, what will be the length, in centimetres, of the model? (3)

Felix then makes a model sailor to stand on the deck.

(b) What is the approximate height of the model sailor? (1)

10.7 A circle has radius 5 cm. Taking π to be 3, estimate:

(a) the circumference of the circle (2)

(b) the area of the circle. (2)

10.8 A circle has radius 6 cm.

Giving your answers to three significant figures, and using either $\pi = 3.142$ or the π value in your calculator, calculate:

(a) the circumference of the circle (2)

(b) the area of the circle. (3)

10.9 The diagram below shows a design made up of a square and four semicircles.

10 cm

Calculate:

(a) the area of the square (1)

(b) the total area of the four semicircles (3)

(c) the area of the design (1)

(d) the perimeter of the design. (2)

10.10 The circumference of the base of a cylindrical tin of dog food is 22 centimetres.

(a) Calculate:

(i) the radius of the base of the tin (3)

(ii) the area of the base of the tin. (2)

The tin is 11 cm tall.

(b) Calculate the volume of the dog food. (Assume that the tin is completely full of dog food.) (2)

John scoops the contents of the tin into Fido's cylindrical bowl, which has a base of diameter 15 cm, and spreads the food so that the surface is level.

(c) Calculate the depth of food in the bowl, giving your answer to the nearest half centimetre. (3)

10.2 Shape

Reflection symmetry in plane shapes

Lines of symmetry, or mirror lines, always go through the centre of a shape. This may help you decide where they are. A shape can have any number of mirror lines.

Only one of these signs does **not** have any reflection symmetry. Which one is it? (Answer at the end of this section.)

Don't forget that mirror lines can go diagonally as well as across or up (think of a square). However, many people think that rectangles and parallelograms have diagonal mirror lines too – do not be caught out!

Rectangles

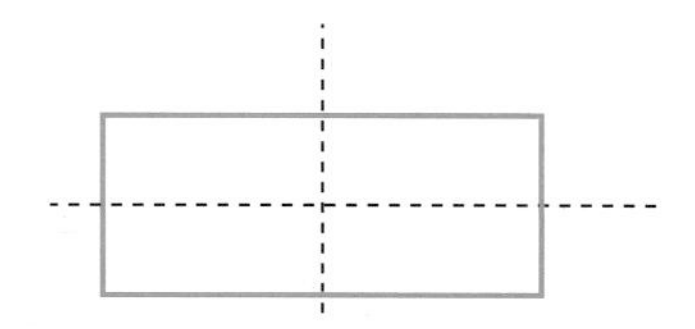

Rectangles have two lines of symmetry.

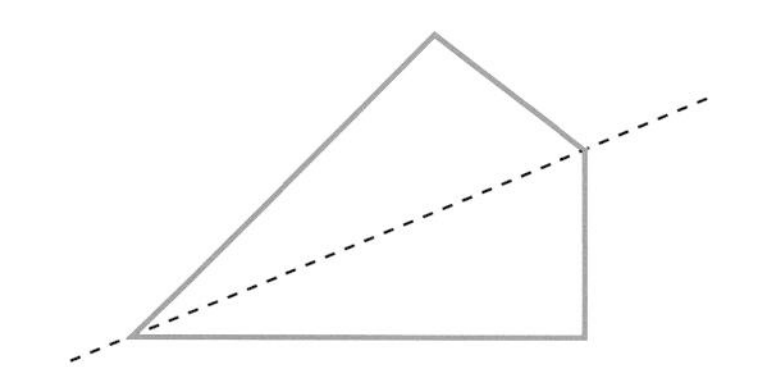

This is what you get if you try to reflect them diagonally! A kite.

Parallelograms

Parallelograms do not have **any** lines of symmetry.

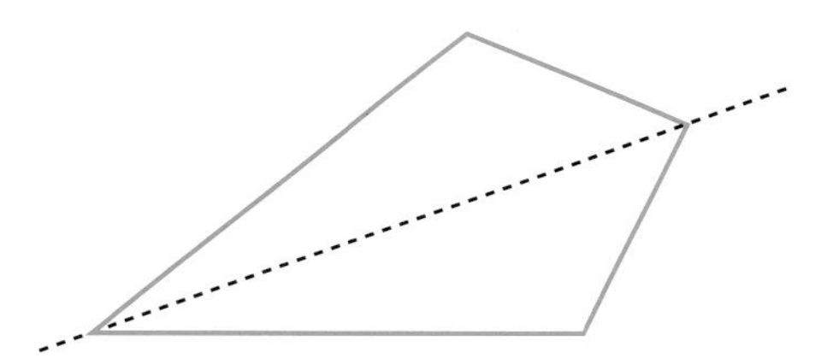

This is what you get if you try to reflect them diagonally. A kite again!

Answer: the second sign has rotation symmetry but not reflection symmetry.

Rotation symmetry in plane shapes

If you can picture in your mind how many right-ways up a picture has, then you know its order of rotational symmetry. Do not worry if you find this difficult to do in your head. It is a very good idea to trace the shape and then turn the tracing paper round to see if the tracing will fit over the original as it turns.

Order of rotation

If a shape has rotation symmetry order 2, then it simply means that the shape looks the same in two positions as it is turned around. Order 3 means it looks the same in three positions and so on.

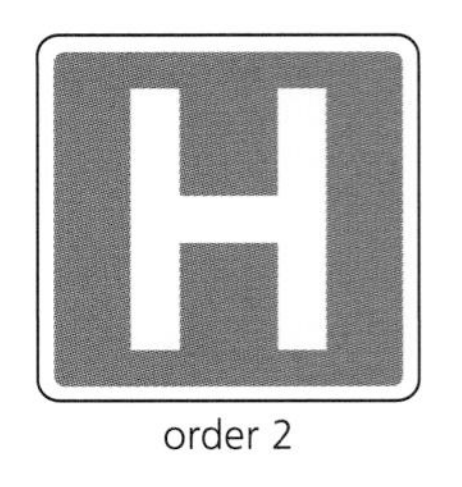
order 2

order 3

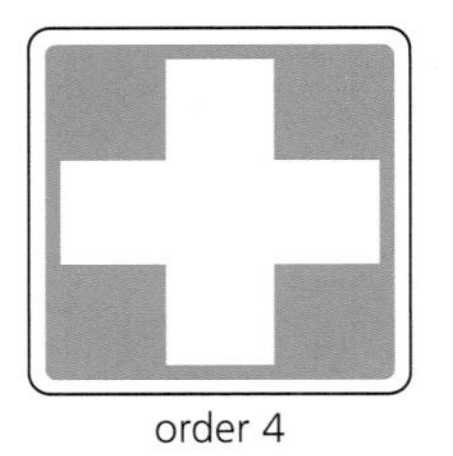
order 4

Centre of rotation

The centre of rotation is the name given to the point on a picture where you would put a pin if you wanted to spin it round. See if you can find the centre of rotation for each of the three previous pictures.

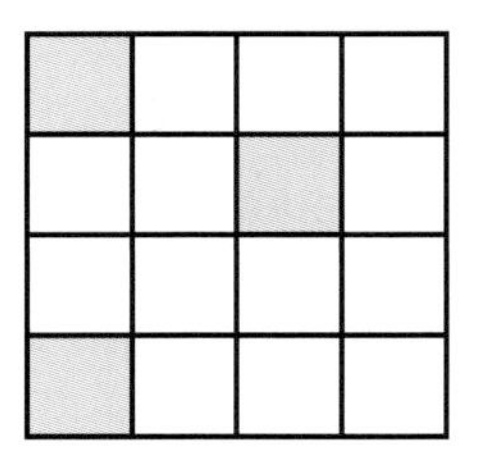

What is the smallest number of squares you have to shade in order to give the pattern a rotation symmetry of order 4?

Answer: 5

Properties of quadrilaterals

Here are the types and properties of the different quadrilaterals.

Diagram of shape	Lines of symmetry	Order of rotation	Notes
square	4	4	The only regular quadrilateral: 4 equal sides and 4 equal angles. Diagonals bisect each other at right angles.
rectangle	2	2	Includes squares (i.e., a square is a special type of rectangle). Non-square ones are called 'oblongs'. Diagonals bisect each other.
rhombus	2	2	Diamond – includes squares. Opposite angles are equal. Diagonals bisect each other at right angles.
parallelogram	0	2	Includes rhombus, rectangle and square. Opposite angles are equal. Diagonals bisect each other.

arrow head (delta)	1	1	The only concave quadrilateral: one diagonal is external. A special type of kite. Diagonals do not intersect.
kite	1	1	Includes rhombus, square and arrowhead. One pair of opposite equal angles. Diagonals intersect at right angles.
(isosceles) trapezium	1 or 0	1	Includes rectangle, parallelogram, and square. The isosceles trapezium has: 2 equal sides and 2 equal angles.

Polygons

Names of the polygons

A **polygon** is a shape with straight sides. In fact, the word is just the Greek for *many angles* (literally *many knees*). Polygons can have any number of sides from three upwards and each polygon has a corresponding special name. Here are the first ten names:

Number of sides	Polygon name
3	Triangle
4	Quadrilateral
5	Pentagon
6	Hexagon
7	Heptagon

Number of sides	Polygon name
8	Octagon
9	Nonagon
10	Decagon
11	Hendecagon*
12	Dodecagon*

*not examined

All candidates should know at least the names of an equilateral triangle, square, hexagon and octagon.

Regular polygon angles

A regular polygon is one in which all the sides are the same length and all the angles are equal. These are made by drawing the required number of points at equal intervals around the circumference of a circle.

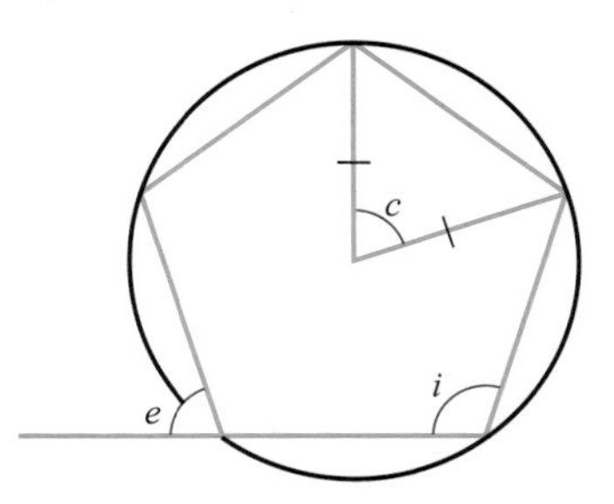

This is a regular pentagon, but the following rules work for any regular polygon.

Let's assume that the polygon has n sides.

(Here $n = 5$)

Then the angle at the centre is given by $c = \frac{360^\circ}{n}$.

The triangle made by joining the centre of the circle to any two vertices is always isosceles. From this fact it is easy to show that the interior angle formula is:

$$i = 180° - c$$
$$= 180° - \frac{360°}{n}$$

Imagine making the polygon out of a straight length of wire. The exterior angle is the angle through which you would bend the wire at each corner.

$$e = 180° - i$$
$$= \frac{360°}{n} = c$$

Finally for any polygon, the sum of all the interior angles is given by $180(n - 2)°$. The sum of all the exterior angles is always 360°.

Drawings of 3D objects on isometric paper

Isometric paper gives you a clever way to draw 3-D shapes in 2-D.

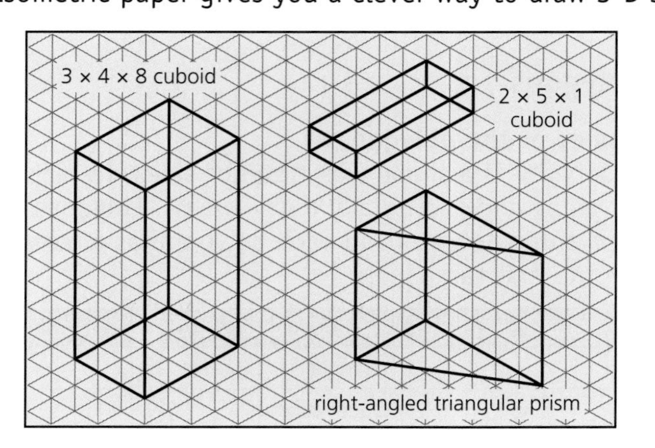

Tips

- Hold the paper the right way up. You need straight vertical lines, not straight horizontal lines.
- Use the lines on the paper to guide your drawing.
- Look at the way right angles are drawn – they are not 90 degrees!
- Use your ruler carefully. Check especially that you are drawing parallel lines of equal length on corresponding edges.

Exam-style questions

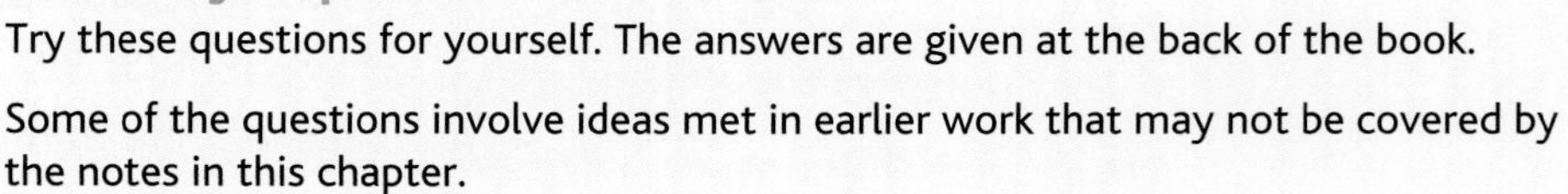

Try these questions for yourself. The answers are given at the back of the book.

Some of the questions involve ideas met in earlier work that may not be covered by the notes in this chapter.

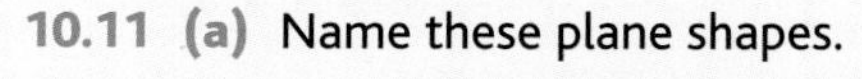

10.11 (a) Name these plane shapes. (3)

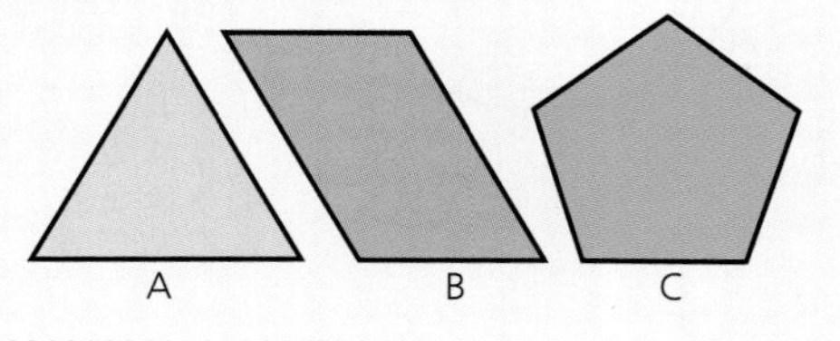

(b) Sketch the following shapes, clearly marking all equal angles and lengths and parallel sides:

(i) rhombus (1)

(ii) kite (1)

(iii) isosceles trapezium (1)

(c) Sketch the following shapes and draw all lines of symmetry:

(i) square (1)

(ii) kite (1)

(iii) regular hexagon (1)

10.12 Describe fully the reflection and/or rotation symmetry of the following plane shapes:

(a) regular hexagon (2)

(b) parallelogram (1)

(c) isosceles triangle (1)

10.13 (a) Name these solid shapes. (3)

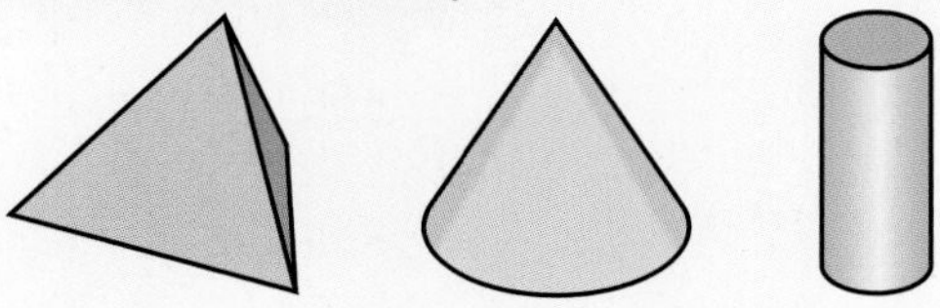

(b) Sketch a **net** for the triangular prism shown below. (1)

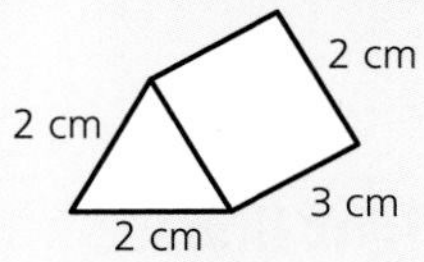

2 cm

2 cm

3 cm

2 cm

10.14 Choose your answers from the following list of quadrilaterals.

square rectangle rhombus parallelogram kite isosceles trapezium

Note: A quadrilateral may appear more than once in your answers.

List the quadrilaterals that have:

(a) all sides the same length (2)

(b) all pairs of opposite angles equal (2)

(c) diagonals intersecting at right angles (2)

(d) diagonals that are equal (2)

(e) no reflection symmetry (2)

(f) no rotation symmetry. (2)

10.15 (a) The exterior angle of a regular hexagon is 60°.

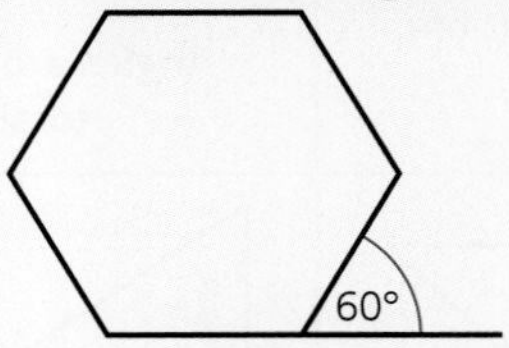

60°

What is the **interior angle** of a regular hexagon? (1)

(b) Bella has drawn a regular **polygon** that has 15 sides.

What size is

(i) the exterior angle (1)

(ii) the interior angle (1)

(iii) the sum of the interior angles? (2)

10.16 On isometric paper, draw representations of the solids that are sketched below:

(a) a cuboid (1)

(b) a step structure (3)

(c) a triangular prism. (3)

10.3 Space

Angle laws for straight lines

Here's a summary of the angle laws for straight lines.

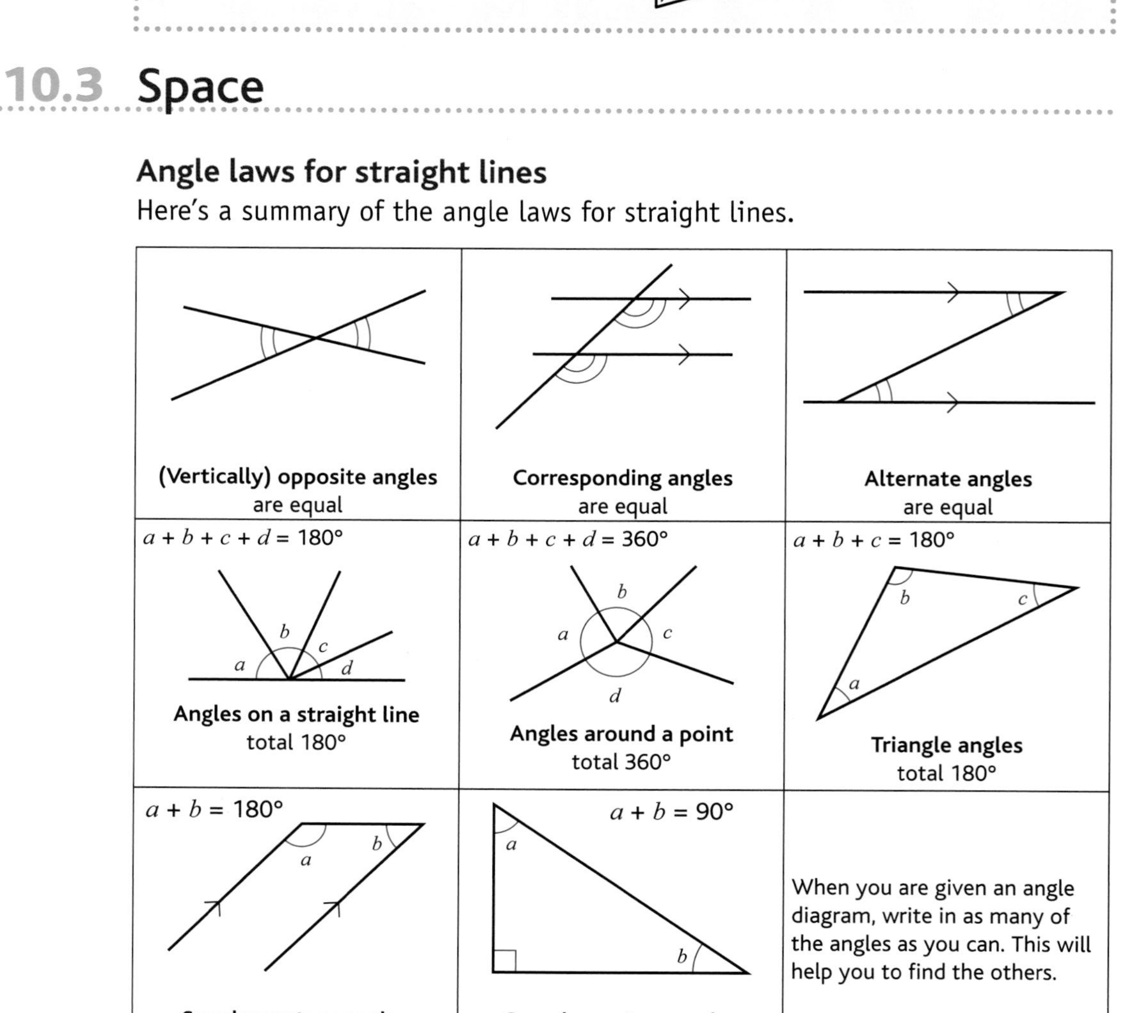

Remember

Complementary or Supplementary? Remember that C comes before S in the alphabet, so C is smaller than S. Also, C stands for Corner and S stands for Straight.

Angle laws for polygons

Level 1
questions will be restricted to solving problems using the angle properties of intersecting and parallel lines.

Most of what you need to know about angles in regular polygons is in Section 10.2. Look out for isosceles triangles! The fact that they have two equal sides and two equal angles can often provide you with lots of clues. All regular polygons (including the equilateral triangle) have both rotation and reflection symmetry. This is always the same as the number of sides. For example, a regular octagon has 8 mirror lines and rotation symmetry of order 8

Ruler and compass constructions

Perpendiculars to a line can be constructed either at a point on the line, or from a point elsewhere.

Constructing a perpendicular at a given point on a line

Step 1 Place the compass point at the point on the line and make an arc on the line either side of the point.

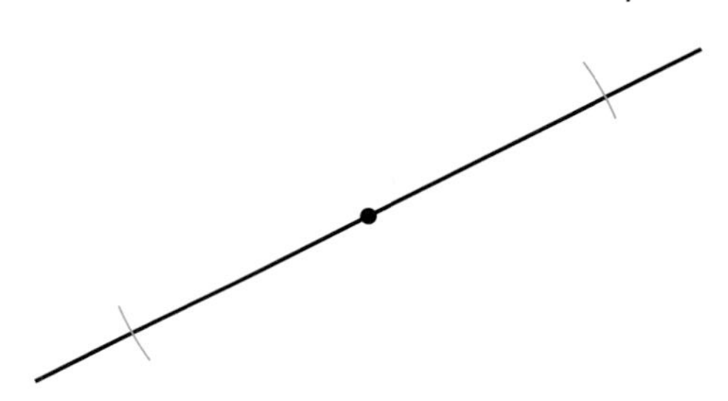

Step 2 Bisect the line segment created between the two arcs (see Chapter 9 section 9.3).

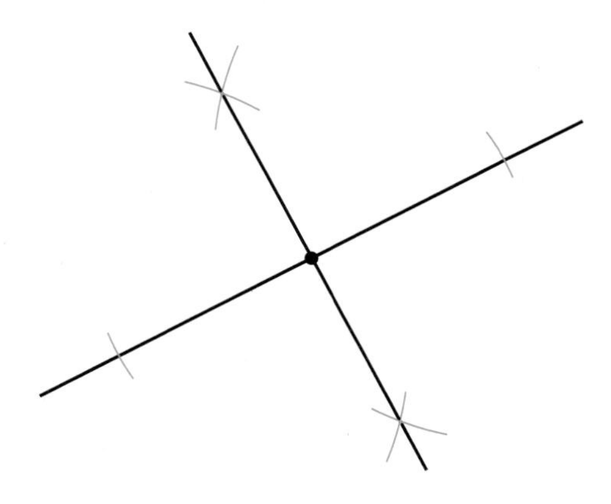

Constructing a perpendicular from a given point to a line

This sounds very similar to the one above, but this time the given point is not on the line.

Step 1 Place the compass point at the given point and make two arcs on the given line.

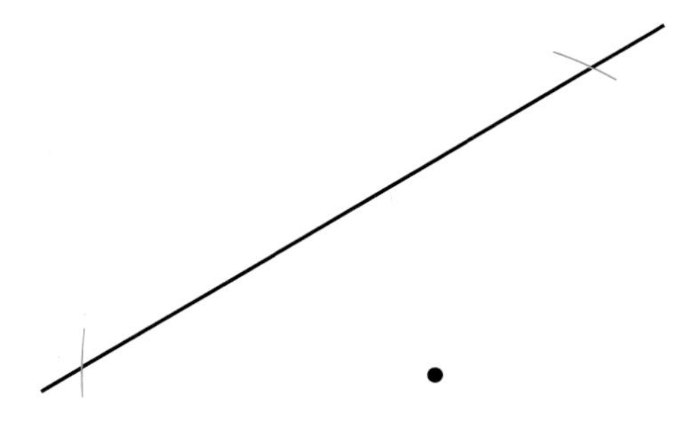

Step 2 Bisect the line segment created between the two arcs as above.

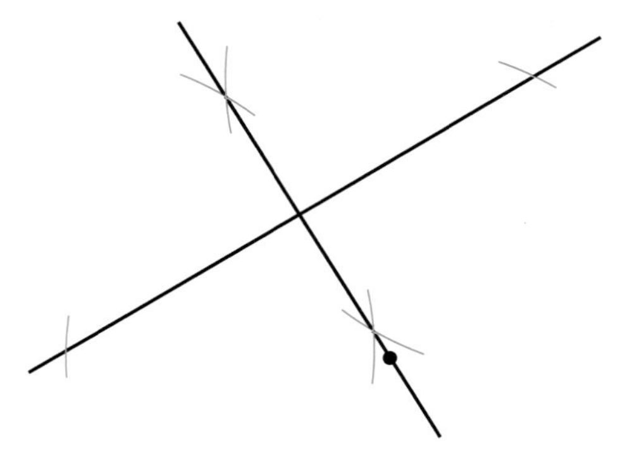

Check that the two lines meet at right angles in each case.

Enlargement and linear scale factor

Here we review the terms **scale factor** and **centre of enlargement**, which are used when making enlargement constructions.

Constructing an enlargement

Step 1 Draw a line from the centre of enlargement to each corner of the shape *and keep going* (it is important that you go beyond the corners).

Step 2 For each corner of the shape, measure the distance from the centre of enlargement and multiply this by the scale factor. Mark a new point along the same line of this new distance from the centre of enlargement.

Step 3 Join up all the points made in Step 2.

Example

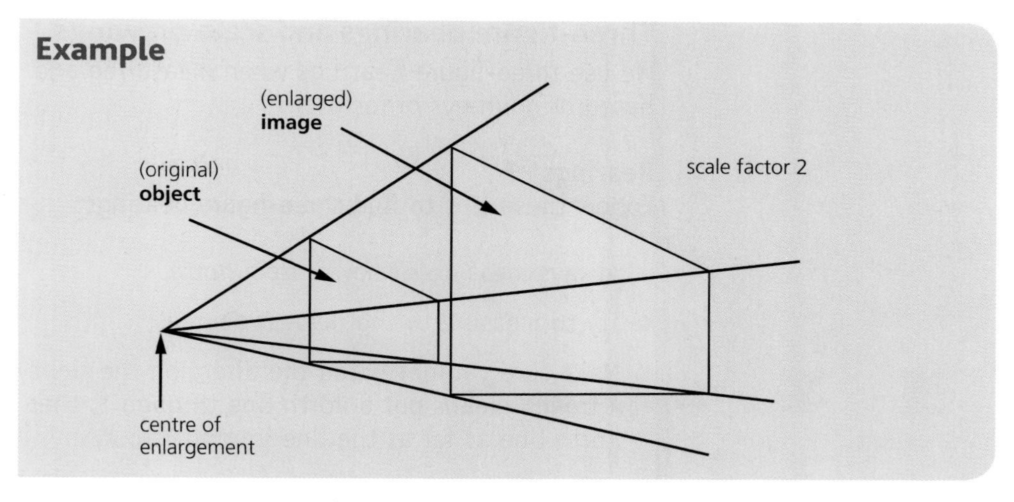

When one shape is an enlargement of the other we call them **similar**.

Finding the centre of enlargement and the scale factor
If the construction has already been made, draw a line from each corner of the image to its corresponding corner on the object *and keep going*. The point where these lines intersect is the centre of enlargement. To find the scale factor, just choose any two corresponding edges: the scale factor is the length of the image edge divided by the length of the object edge.

Area factor for enlargements

Recall from the previous section that the scale factor was the number by which all lengths were multiplied to make the enlargement (so a scale factor of 2 produced a ×2 enlargement).

Area factor is similar. It is the number by which the original area is multiplied to obtain the area of the enlargement. The clever part of it is that there is an easy link between the two.

To find the area factor you just square the scale factor.

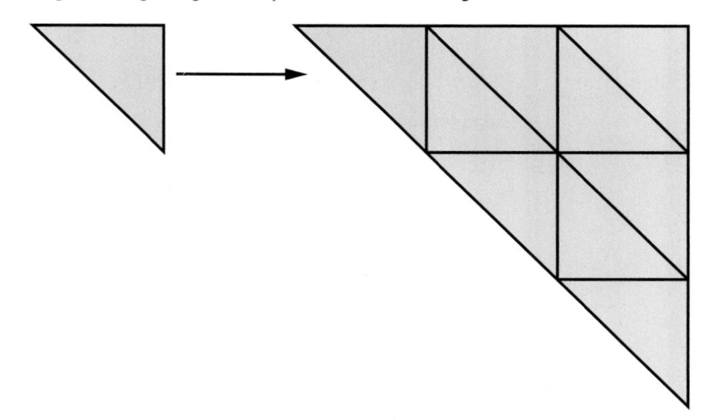

The scale factor is 3, so the area factor is $3 \times 3 = 9$

We can see that the original shape fits exactly nine times into the enlargement.

Similarly for this enlargement (scale factor 2).

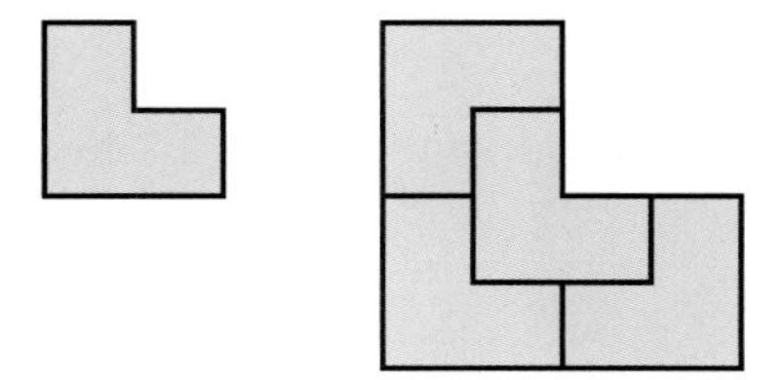

It is true for every shape!

Three-figure bearings and scale drawings

We use three-figure **bearings** when measuring and drawing constructions based on journeys or location.

Bearings

Follow these tips to find three-figure bearings:

- Always measure *clockwise* from *north*.
- Try to measure to the nearest degree.
- Make sure you have read the direction the right way round. The bearing of X from Y means put a north line through Y, then measure round from the north line as far as the line joining Y to X.

Examples

Find the bearing of:

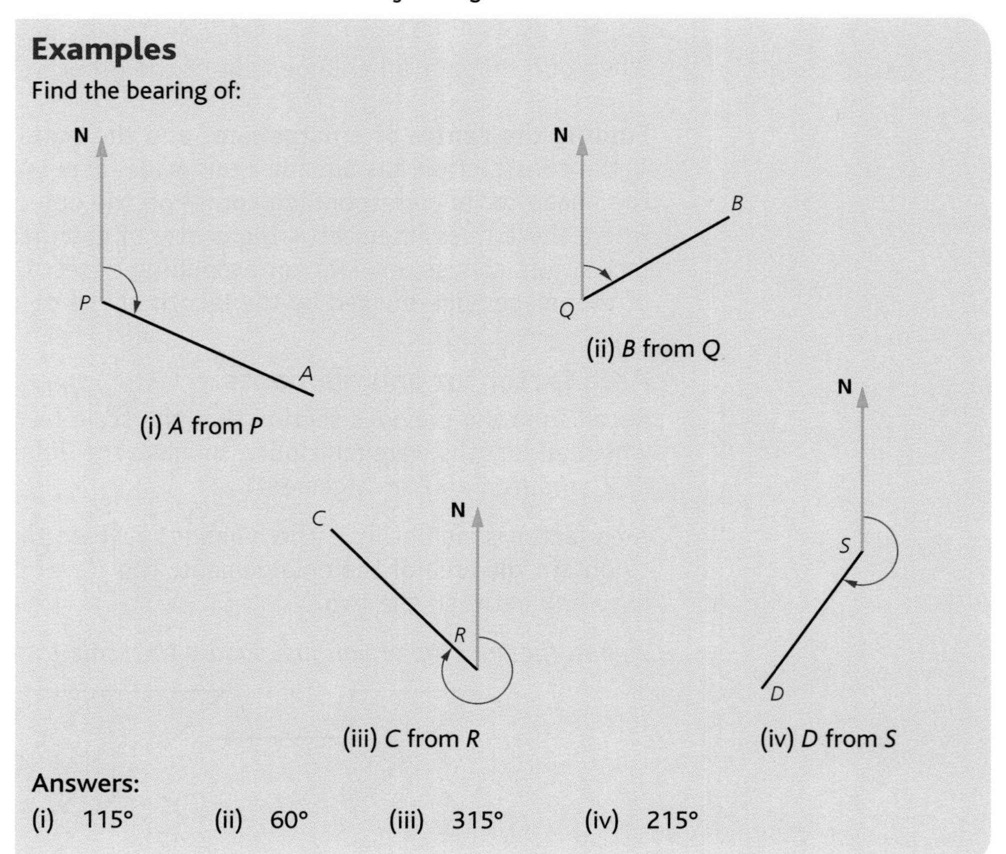

Answers:

(i) 115° (ii) 60° (iii) 315° (iv) 215°

Scale drawings

We can use scale drawings to find the answers to problems by measuring.

Example

An amateur pilot leaves the airfield in his plane and flies towards an island that is 10 km away on a bearing of 095°. However, after he has flown 10 km, he notices he has been flying on a bearing of 085° by mistake. How far is he from the island?

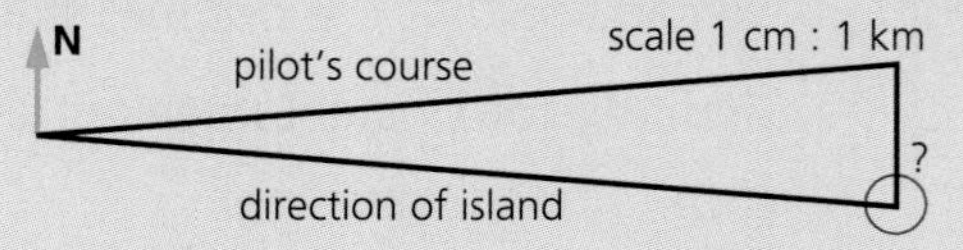

Careful measuring gives the length of the line as 1.7 cm, and so the pilot is 1.7 km or 1700 m off course!

Level 3 ■ Pythagoras' theorem

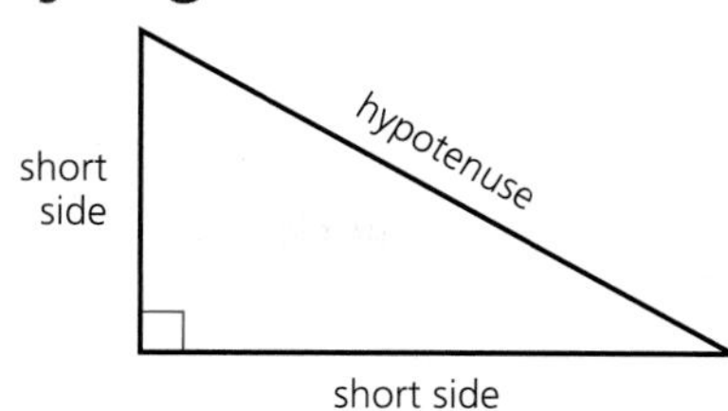

Pythagoras' theorem tells us about the lengths of the sides in right-angled triangles.

'The square on the hypotenuse is equal to the sum of the squares on the other two sides.'

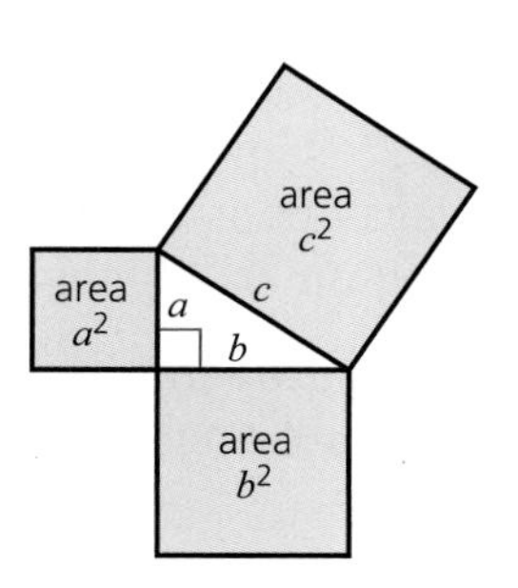

If the hypotenuse is labelled c and the two shorter sides are labelled a and b respectively, then Pythagoras' theorem says:

$$a^2 + b^2 = c^2$$

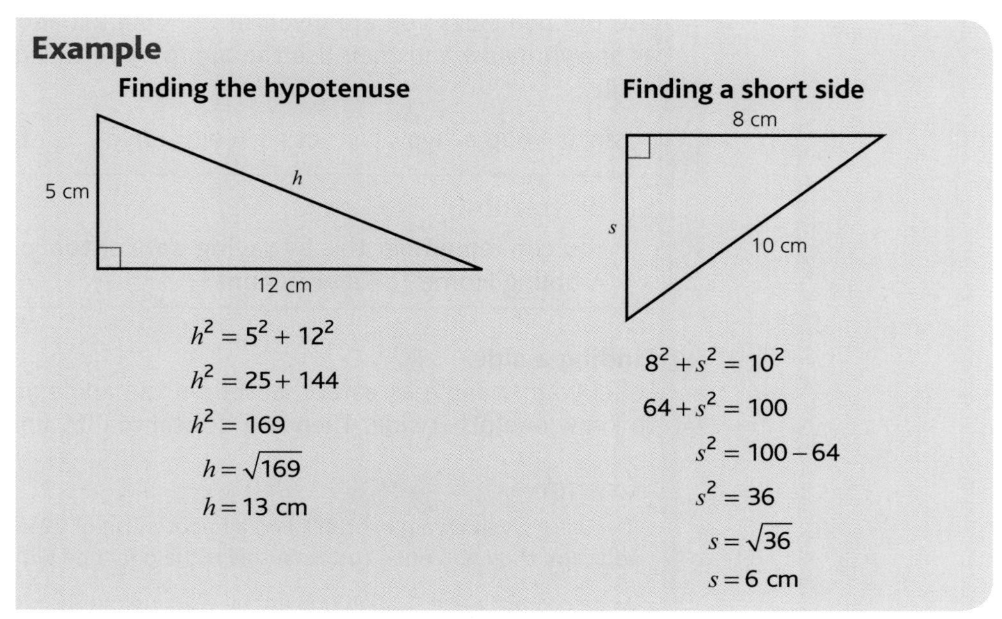

Example

Finding the hypotenuse

$h^2 = 5^2 + 12^2$

$h^2 = 25 + 144$

$h^2 = 169$

$h = \sqrt{169}$

$h = 13$ cm

Finding a short side

$8^2 + s^2 = 10^2$

$64 + s^2 = 100$

$s^2 = 100 - 64$

$s^2 = 36$

$s = \sqrt{36}$

$s = 6$ cm

Trigonometry

This topic is not included in the ISEB Common Entrance syllabus, but it is in KS3.

Some of you may start to study trigonometry as an extension topic, so brief notes are included here.

When we know the length of two sides of a right-angled triangle, we can use Pythagoras' theorem to find the length of the third side (previous section).

When we know the length of only one side, but we also know the size of one of the acute angles, we can use trigonometry.

Trigonometry uses **sine**, **cosine** and **tangent**. Your calculator should have three buttons labelled **sin**, **cos** and **tan**.

Above these buttons you usually see $\sin^{-1}$, $\cos^{-1}$ and $\tan^{-1}$, and these just do the opposite or inverse function. You also need to check your calculator shows **d** or **deg** and not **r** (**radians**) or **g** (**grads**) so that it works out the angles in degrees. (See your Maths teacher if you need to change this – it usually involves pressing MODE or something similar.)

Triangle anatomy

There are special names for the parts of a right-angled triangle.

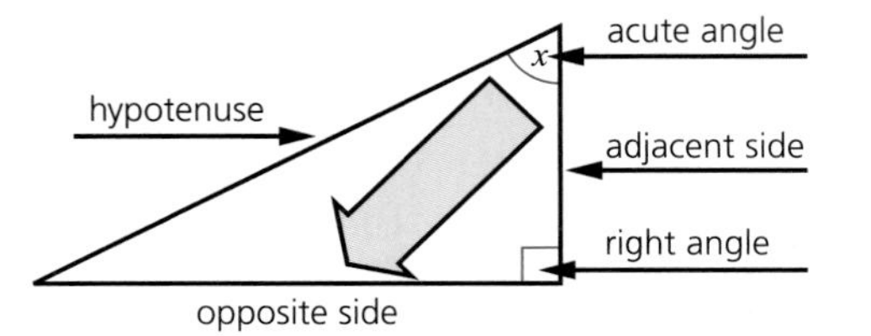

- The **hypotenuse** (hyp) is always the longest side, and always opposite the right angle.
- The **opposite side** (opp) is always opposite the acute angle you are studying (imagine a torch shining on the opposite wall).
- The **adjacent side** (adj) is always between the acute angle and the right angle, adjacent to (next to) the acute angle.

Finding an angle

Use the two sides you are given in the diagram above, divide one by the other as shown below and then use the appropriate inverse function to find the angle.

sin x = **o**pp ÷ **h**yp **c**os x = **a**dj ÷ **h**yp **t**an x = **o**pp ÷ **a**dj

Remember

You can remember this by saying '**sohcahtoa**' or 'Sir Oliver's Horse Came Ambling Home To Oliver's Aunt'!

Finding a side

Label your triangle as earlier, based on the angle you are given. You only need to know one other side. Then use 'sohcahtoa' to find the missing side.

Example

If you are given an angle of 42° and a hypotenuse of 12 cm, and you have to find the adjacent, then you need cos since this is the only one with adj and hyp.

Write cos 42° = ? ÷ 12 and find adj by working out 12 × cos 42° = 8.92 cm.

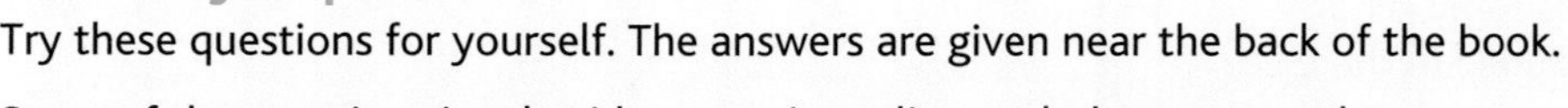

Exam-style questions

Try these questions for yourself. The answers are given near the back of the book.

Some of the questions involve ideas met in earlier work that may not be covered by the notes in this chapter.

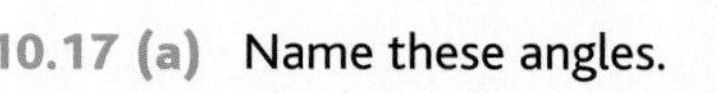
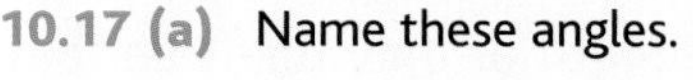
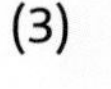

10.17 (a) Name these angles. (3)

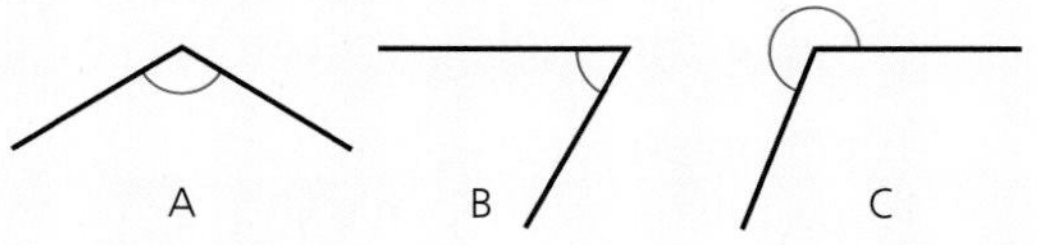

(b) On these diagrams, clearly mark all angles that are the same as those indicated with stars. (3)

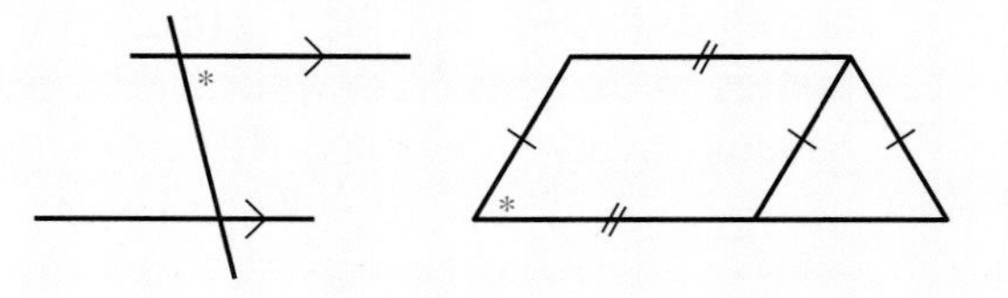

10.18 Look at this diagram.

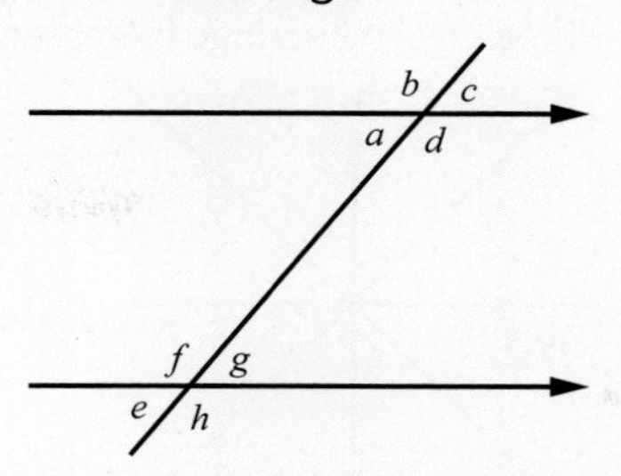

(a) Name an angle that is **alternate** to angle a. (1)

(b) Name an angle that is **vertically opposite** to angle b. (1)

(c) Name an angle that is **corresponding** to angle c. (1)

10.19 Look at this diagram.

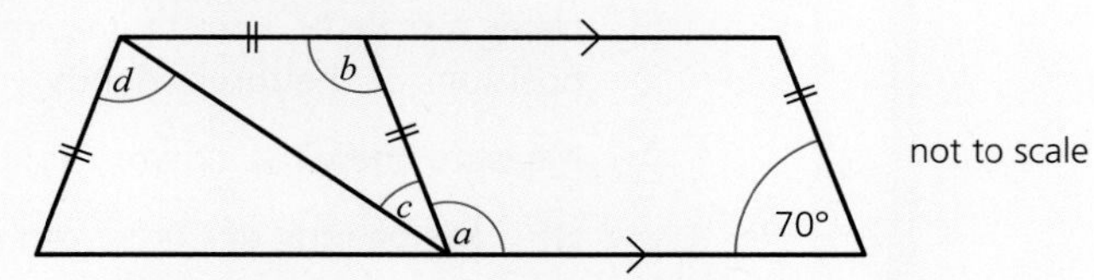

(a) Calculate the angles marked a and b. (2)

(b) Calculate the angles marked c and d. (4)

10.20 Look at this diagram.

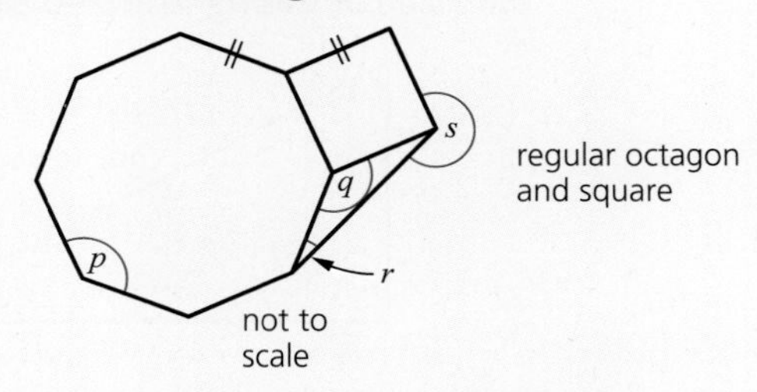

(a) Calculate the angles marked p and q. (2)

(b) Calculate the angles marked r and s. (4)

10.21 (a) Enlarge triangle A using (1, 1) as the centre of enlargement and scale factor 2 (2)

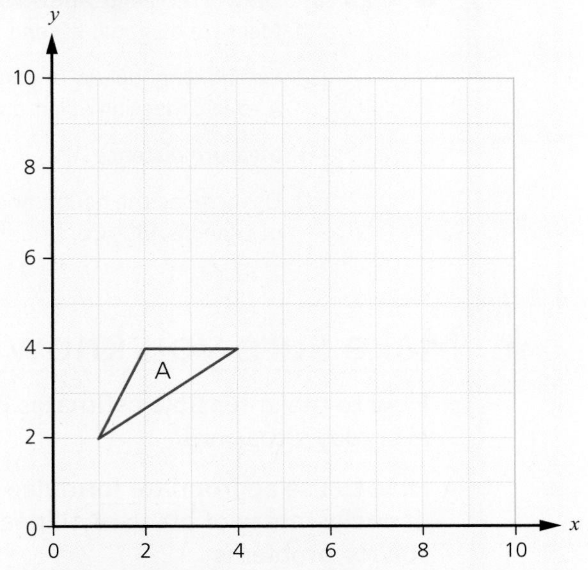

(b) How many times the area of triangle A is the area of the enlargement? (1)

10.22 (a) Complete this diagram showing the points of the compass and the three-figure bearings in degrees. (4)

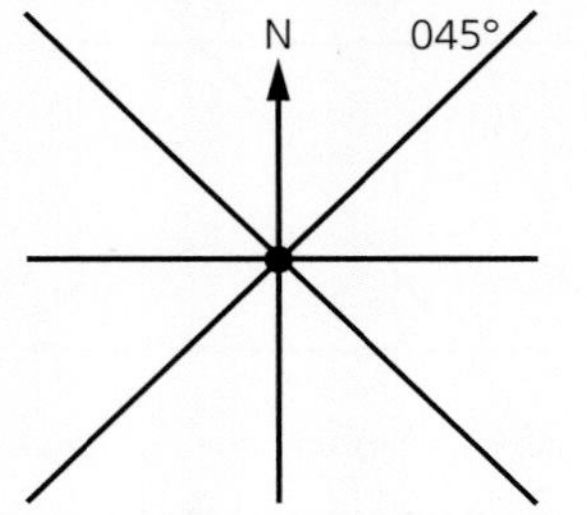

(b) Yoko and Zac are standing in a field. Zac is NW of Yoko. What are the compass direction and bearing of Yoko from Zac? (2)

10.23 Alan is standing in the middle of a playing field. Beatrice is 20 metres north of Alan and Chris is 15 metres east of Beatrice.

(a) Using a scale of 1 cm to 2 m, make a scale drawing showing the positions of the three friends. (3)

(b) Measure and write down:

(i) the bearing of Chris from Alan (1)

(ii) the bearing of Alan from Chris (2)

(iii) the distance of Alan from Chris. (2)

Level 3

■ **10.24 (a)** Using Pythagoras' theorem, calculate the length of the third side of this right-angled triangle. (2)

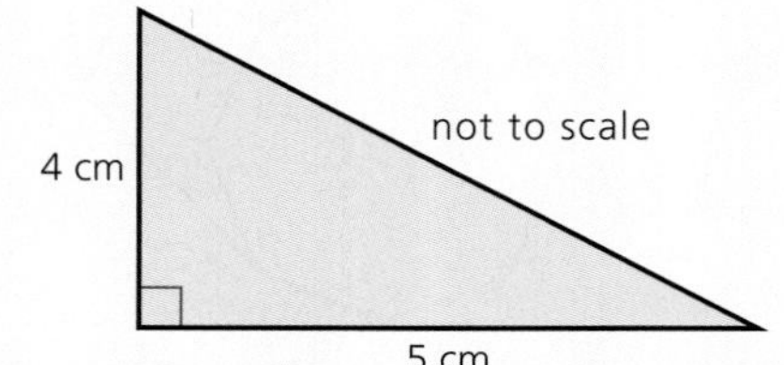

(b) Calculate the length of the diagonal of a rectangle measuring 6 cm by 4 cm. (2)

(c) The diagonals of a square are 8 cm long. What is the length of a side of the square? (3)

■ **10.25 (a)** Draw a rectangle $ABCD$ with AB = 12 cm and AD = 5 cm. Measure diagonal BD and verify this by a suitable calculation. (2)

(b) Extend diagonal BD beyond D. Construct a perpendicular from A to BD, meeting BD at point E. (2)

(c) Measure distance AE. (1)

(d) Using AE as the height and BD as the base, calculate the area of triangle ABD. Comment on your answer. (2)

★ Make sure you know

★ How to make sensible estimates of a range of measures in relation to everyday situations

★ How to use appropriate formulae for finding circumferences and areas of circles, areas of plane rectilinear figures and volumes of cuboids when solving problems

★ How to calculate lengths, areas and volumes in plane shapes and right prisms

★ How to use compound measures, such as speed

★ How to identify all the symmetries of 2-D shapes

★ How to use the properties of quadrilaterals in classifying different types of quadrilateral

★ How to recognise and use common 2-D representations of 3-D objects

★ The angle sum of a triangle and the sum of angles at a point

★ How to enlarge shapes by a positive whole-number scale factor

★ How to use three-figure bearings to define direction

★ How to solve problems using angle properties of intersecting and parallel lines, and explain these properties

★ How to solve problems using angle and symmetry properties of polygons and explain these properties

★ How to construct the perpendicular to a line through any point on the line or through any point not on the line

Level 3 ■ ★ How to apply Pythagoras' theorem when solving problems in two dimensions

★ How to use the glossary at the back of the book for definitions of key words

Levels 1 and 2 require candidates to find the volume of a cuboid, the areas of triangles, parallelograms, trapezia and circles, and the perimeter or circumference of these shapes. The necessary formulae will be given at Level 1. **Level 3** additionally requires candidates to find the volume of a prism, and the radius or diameter of a circle, given the circumference or area.

Speed is the only compound measure that will be tested. **Level 1** candidates might be expected to answer a question such as: 'If Jane takes 2 hours to cycle 30 km, what is her average speed?' **Level 2** candidates might be expected to answer a question such as: 'How far will Tom travel if he cycles for three quarters of an hour at a speed of 16 kilometres per hour?' **Level 3** candidates might be expected to calculate the average speeds of multi-stage journeys.

Test yourself

Before moving on to the next chapter, make sure you can answer the following questions.

The answers are near the back of the book.

1 (a) Estimate the diameter of

(i) a hockey ball

(ii) a table-tennis ball

(iii) a basketball.

(b) Estimate the mass of

(i) a hockey ball

(ii) a table-tennis ball

(iii) a banana.

(c) Estimate the capacity of

  (i) an ordinary mug

  (ii) a standard bath.

2 The diagram shows a Chinese cash coin that is circular with a square hole in the middle. The diameter of the coin is 29 mm and the square hole has side 6 mm.

(a) Calculate the area of one side of the coin.

The thickness of the coin is 2 mm.

(b) Calculate the volume of the metal in the coin.

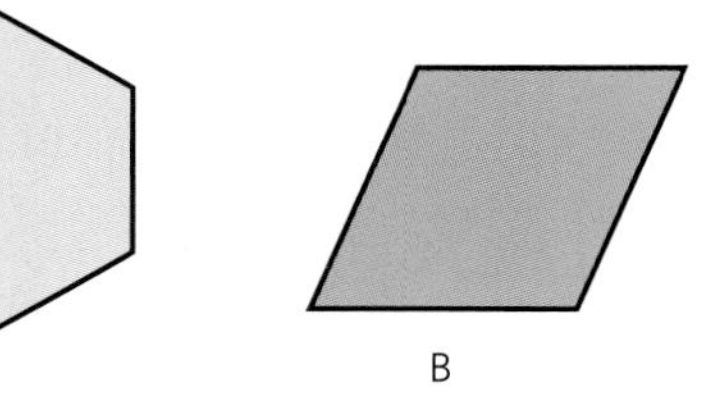

3 In a school swimming competition, the individual swimmers' times in the 4 × 200 metre freestyle relay are:

swimmer A 2 minutes 5 seconds

swimmer B 1 minutes 59 seconds

swimmer C 2 minutes 11 seconds

swimmer D 2 minutes 9 seconds

(a) What was the total time for the team?

(b) What was the mean (average) time for a swimmer?

(c) By how many seconds did the fastest swimmer beat the mean time?

4 Describe in full the symmetries of shapes A and B.

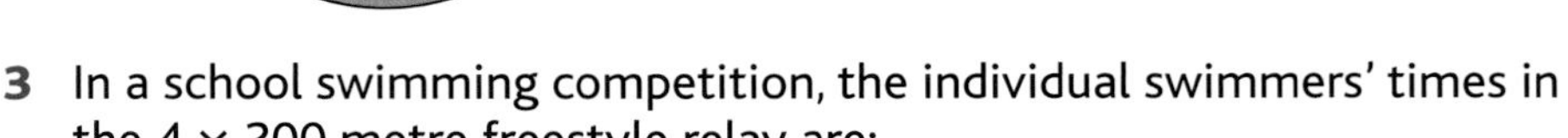

5 Calculate the angles a, b, c, d and e.

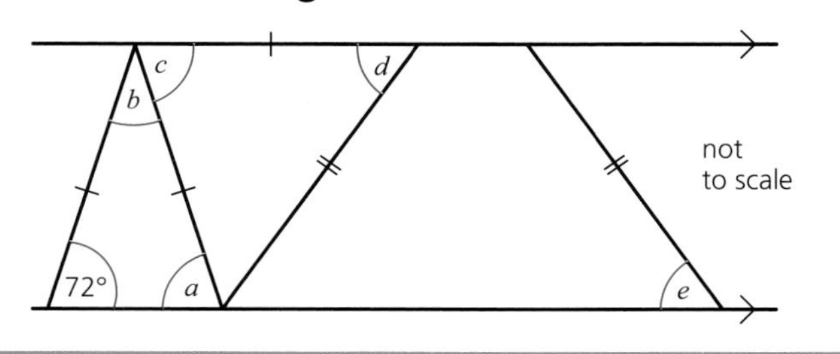

6 Prepare a co-ordinate grid with axes numbered from 0 to 10

(a) Plot the points (1, 1), (2, 1) and (2, 3) and join the points to form triangle A.

(b) **Reflect** triangle A in the line $x = 3$ and label the image B.

(c) **Rotate** triangle A through 90° clockwise about (2, 3) and label the image C.

(d) **Translate** triangle A six units to the right and three units up. Label the image D.

(e) **(i)** **Enlarge** triangle A, with (0, 0) as the centre of enlargement and scale factor 3, to form triangle E.

(ii) How many triangles congruent to triangle A could fit into triangle E?

7 The diagram shows a solid made from interlocking centimetre cubes, drawn on an isometric dotted grid.

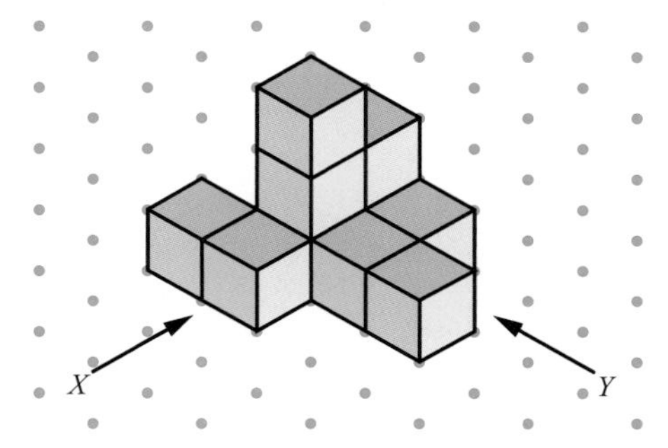

Draw views of the solid seen from positions X and Y.

8 Fiona, Grant and Henrietta are standing on a hockey pitch. Fiona is 20 m due west of Grant. Henrietta is 35 m south-east of Fiona.

(a) Make a scale drawing to show the relative positions of the three children. It is always a good idea to make a sketch first.

(b) Use your scale drawing to find

(i) the bearing of Grant from Henrietta

(ii) the distance of Grant from Henrietta.

Level 3

■**9** The diagonals of a rhombus are of lengths 16 cm and 12 cm.

(a) Use Pythagoras' theorem to find the length of a side of the rhombus.

The rhombus is cut along its diagonals to form four congruent triangles. These four triangles are rearranged to form a rectangle.

(b) Draw labelled sketches to show the two different shaped rectangles that could be formed.

(c) What is the length of a diagonal of the rectangle with the larger perimeter?

Statistics and probability (1)

11.1 Statistics

Sorting shapes; Venn and Carroll diagrams

In this section we use a flow chart to sort six different quadrilaterals (can you name them all?) and then we sort them, in a different way, using a Carroll diagram.

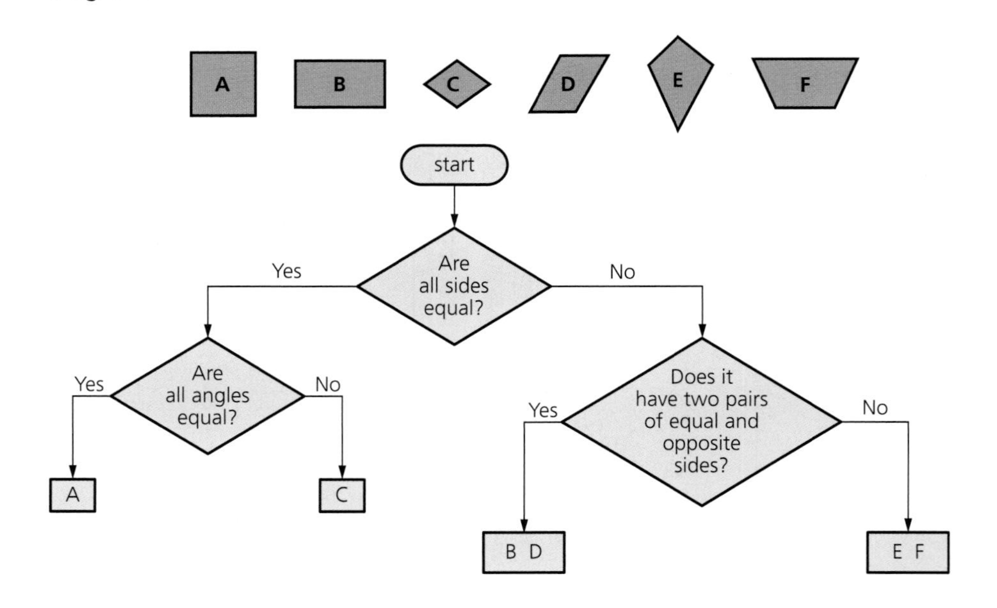

Carroll diagram

	Sides all equal	Sides not all equal
Angles not all equal	C	D E F
Angles all equal	A	B

Finally we use a Venn diagram to sort the shapes. Which way do you prefer to sort the quadrilaterals?

Venn diagram

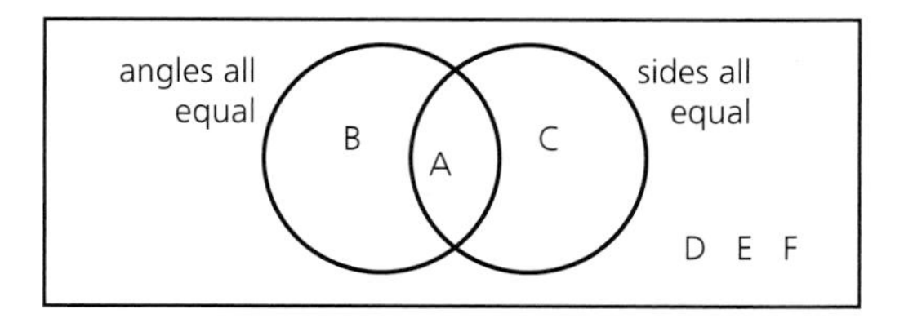

Frequency tables using discrete data

Discrete data are the sort of data you can count or sort into easy categories. For example, we could ask:

'How many children are in your family?' or 'What is your favourite colour?'

A **frequency table** is a good way of collecting and organising discrete data. In this example, we look at the work of a pupil carrying out a survey on favourite breakfast cereals amongst people:

Cereal	Tally	Frequency
Corn flakes	𝍸 I	
Wheat flakes	III	
Bran flakes	I	
Oat flakes	IIII	

At this stage she had only asked 14 people. Eventually, however, it looked like this:

Cereal	Tally	Frequency
Corn flakes	𝍸 𝍸 II	12
Wheat flakes	𝍸 𝍸	10
Bran flakes	IIII	4
Oat flakes	𝍸 IIII	9
	Total	35

The pupil has now filled in the frequencies (row totals) and found the overall total. The table is complete.

Finding the mode, median and range

When we have lots of data (information), usually after a survey, we need a way of understanding the numbers and results more clearly. Maths to the rescue! The branch of mathematics that we use for this task is called statistics; it is a very useful topic to study.

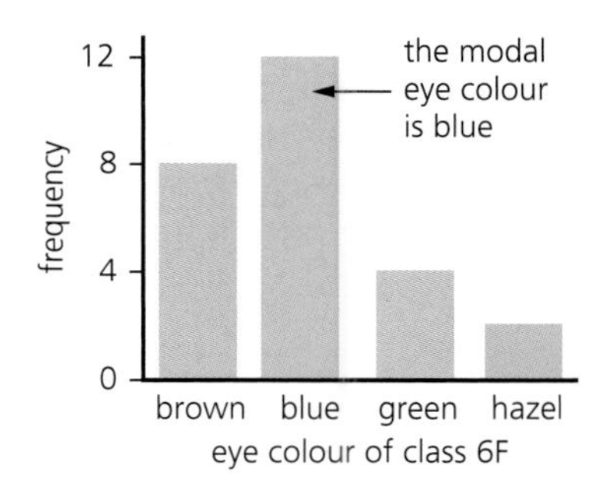

The **mode** is the category in the survey that came up most often, i.e. the one with the greatest frequency. The mode is the most common, or most popular choice.

The **median** means *middle value*. If we sort our numerical data into order, the median is the value in the middle, or halfway between the middle two, if the number of items of data is even.

The **range** tells us how spread out numerical results are. It is simply the difference between the smallest and largest values in the data.

Example

I put the question 'How many children are in your family?' to 41 different families, and received these answers (sorted into increasing order of size):

1, 1, 1, 1, 1, 1, 1, 1, 2, 2, 2, 2, 2, 2, 2, 2, 2, 2, 2, (3), 3, 3, 3, 3, 3, 3, 3, 3, 3, 4, 4, 4, 4, 4, 4, 4, 5, 5, 5, 6

So the mode is 2 (most common), the median is 3 (half way through the list), and the range is 5 (since 6 − 1 = 5).

Calculating the mean; comparing two distributions

How to find the mean

This is one of the simplest calculations in statistics. All you do is add up your data and divide by the number of items in the list.

Example

Find the mean of the following pupil masses (all in kg):

42, 37, 39, 40, 45, 37, 35, 49, 36, 43

Add them up.

42 + 37 + 39 + 40 + 45 + 37 + 35 + 49 + 36 + 43 = 403

There are ten pupils in this set of data so we divide by 10

403 ÷ 10 = 40.3

Thus the mean pupil mass is 40.3 kg.

Comparing two distributions

When comparing two distributions, we should take the mean, mode, median and range into consideration.

Example

A cricket team needs 25 runs in order to win. A choice has to be made between two players whose past run totals look like this:

Arthur	19, 24, 30, 27, 23, 24, 29, 21, 31, 25		
	mean 25.3	median 24.5	range 12

Daley	8, 48, 22, 32, 19, 42, 7, 59, 0, 18		
	mean 25.5	median 20.5	range 59

Note: for Arthur, the median is halfway between 24 and 25; for Daley, the median is halfway between 19 and 22

So whom should they choose? Daley certainly has a greater mean, but he is very inconsistent (large range) and the median tells us that over half his previous games fell short of the desired 25 runs. Arthur, on the other hand, although he has a lower mean, is a much more consistent player (small range), and the median shows us that he had as many results above 25 as below it.

Whom would *you* choose?

Grouped data and equal class intervals

Using grouped data

Junior Splash Swimming Club recently asked its 55 members how many lengths they swam last weekend. The raw data results were as follows:

14	31	54	40	22	45	57	26	33	56	52
34	24	37	33	36	27	36	34	23	60	29
47	47	57	21	17	37	22	11	32	23	44
32	25	41	42	38	29	44	25	38	39	23
43	46	50	36	38	26	44	33	33	52	34

To make it easier to follow, we group the data:

Number of lengths	Tally	Frequency
0 to 9		0
10 to 19	III	3
20 to 29	𝍸 𝍸 IIII	14
30 to 39	𝍸 𝍸 𝍸 IIII	19
40 to 49	𝍸 𝍸 I	11
50 to 59	𝍸 II	7
60 to 69	I	1
	Total	55

Note the equal **class intervals** (10 lengths each). It is now much clearer that the modal class is 30 to 39 lengths.

From the original data, the range is 49 (60 – 11) lengths. We cannot find the median of **grouped data**, but we can see that it will be somewhere in the 30 to 39 lengths group. In this case, there are 17 who swam fewer than 30 lengths and 19 who swam more than 39 lengths. If we wanted to find it more accurately, we would have to sort the original data.

Line graphs and conversion graphs

Temperature and other time graphs

One type of line graph consists of short line segments joined up to show how something is changing over a period of time. Here is a temperature graph:

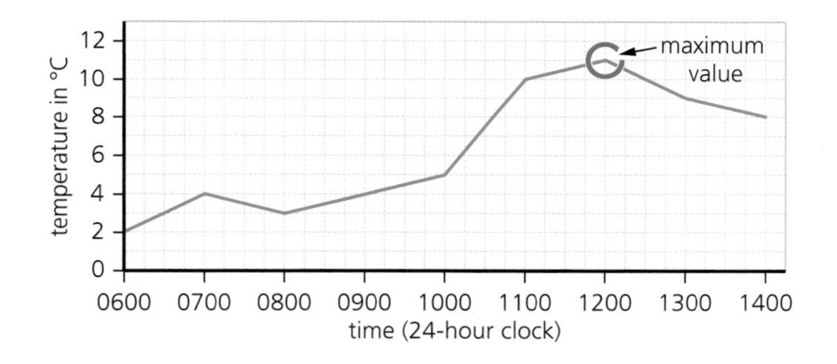

The temperature was measured and plotted every hour, on the hour. The line segments joining these points are really only guesses at the intermediate values – the real change is likely to have been a smooth **curve** throughout the day, but the graph is still a useful way of displaying the collected information.

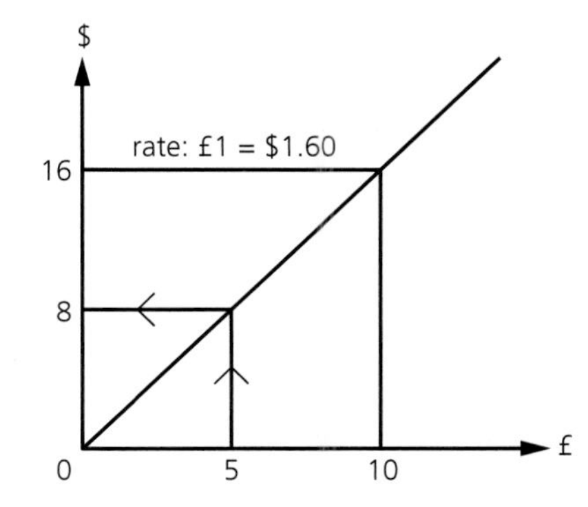

Conversion graphs

Another type of graph is the **conversion graph**. This one converts between pounds (£) and dollars ($). Check that £5 = $8

When drawing conversion graphs, it is important to choose a suitable scale. Choose one that is big enough to read accurately, but do use the squares on the graph paper sensibly.

And do not forget to label the axes!

Exam-style questions

Try these questions for yourself. The answers are given near the back of the book.

Some of the questions involve ideas met in earlier work that may not be covered by the notes in this chapter.

11.1 The first nine counting numbers are:

1 2 3 4 5 6 7 8 9

(a) Write each of these numbers in the correct region of this Carroll diagram. (4)

	odd	not odd
not prime		
prime		

(b) Write each number in the correct region of this Venn diagram. (4)

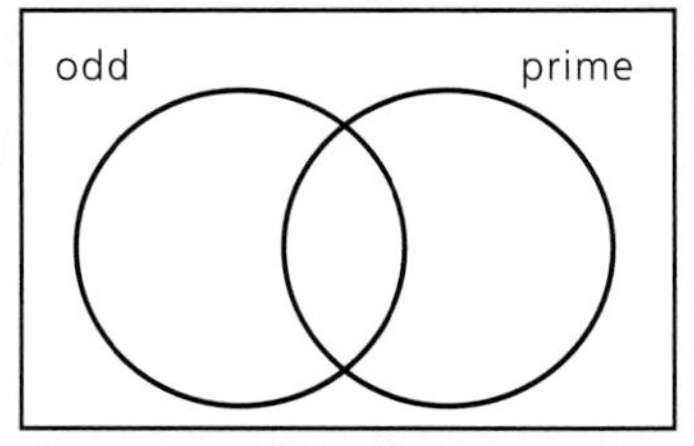

11.2 This pictogram shows the number of Valentine's Day cards received by five friends.

Amy	♥ ♥ ♥ ♥ ♥
Ben	♥ ♥ ♥ ♥ ♥ ♥ ♥
Clare	♥ ♥ ♥ ♥ ♥ ♥ ♥ ♥ ♥ ♥
Dan	♥ ♥ ♥ ♥ ♥ ♥
Enid	♥ ♥ ♥ ♥ ♥ ♥ ♥ ♥

Key: ♥ represents 1 card

(a) How many cards did Ben receive? (1)

(b) What is the total number of cards received by the friends? (2)

11.3 Alison has drawn this graph to show where the children in her school spent their holiday.

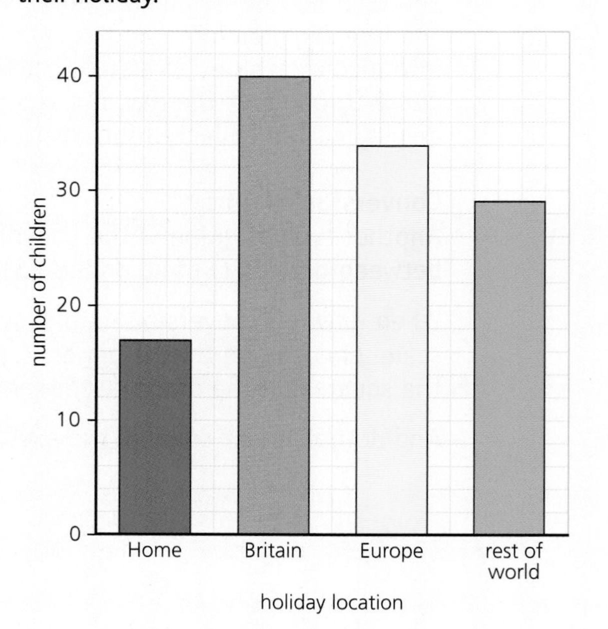

(a) How many children stayed at home? (1)

(b) How many children took part in Alison's survey? (3)

(c) How many more children went to Europe than went to the rest of the world? (1)

(d) What fraction of the children went away from home on holiday in Britain? (2)

11.4 Here are the numbers of books read in a week by five children.

Fay	Gill	Hal	Ian	Jon
1	2	7	1	4

For these numbers, what is

(a) the range (1)

(b) the median (1)

(c) the mode (1)

(d) the mean? (2)

11.5 The masses of five children in a family are 40 kg, 38 kg, 32 kg, 41 kg and 39 kg.

What is the mean mass of the children? (3)

11.6 The children in Miss de Cova's class have made this list of their heights in metres:

1.41 1.52 1.28 1.43 1.49

1.48 1.37 1.52 1.39 1.47

1.30 1.51 1.40 1.48 1.57

1.45 1.33 1.42 1.44 1.45

(a) Complete this table. (4)

Height (m)	Tally marks	Frequency
1.25 to 1.29		
1.30 to 1.34		
1.35 to 1.39		
1.40 to 1.44		
1.45 to 1.49		
1.50 to 1.54		
1.55 to 1.59		
	Total	

(b) Use the table to draw a frequency diagram. (4)

(c) Which group of heights is the mode? (1)

11.7 This line graph shows the amount of fuel in the fuel tank of Mr McPhee's car.

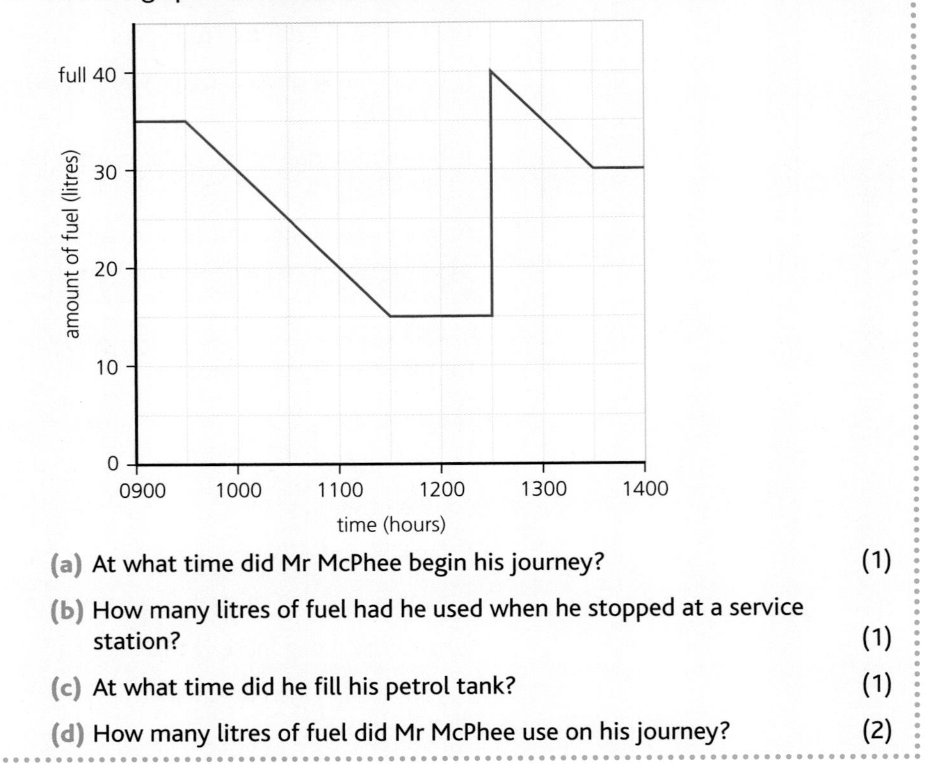

(a) At what time did Mr McPhee begin his journey? (1)

(b) How many litres of fuel had he used when he stopped at a service station? (1)

(c) At what time did he fill his petrol tank? (1)

(d) How many litres of fuel did Mr McPhee use on his journey? (2)

11.2 Probability

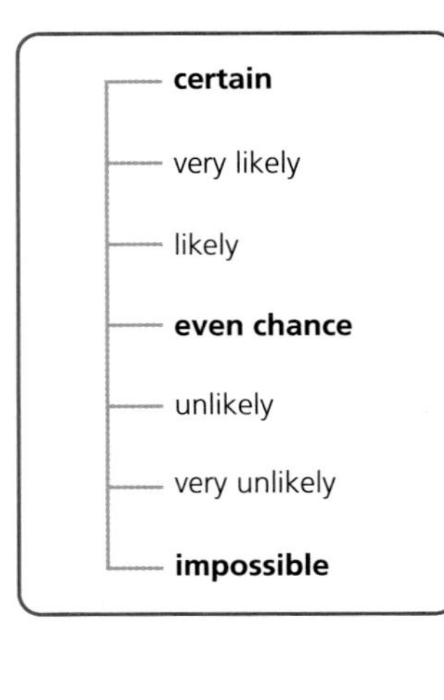

Words to describe probability

How likely?

How likely is it that the sun will rise tomorrow morning? **Certain.**

How likely is it that your pet stick insect will grow into a tree? **Impossible**!

In-between questions are more difficult. How likely is it that it will rain tomorrow? It is difficult to say, even if you have had rain every day so far this week. The weather could improve!

Even so, we can use this word scale to describe and compare the probabilities (chance or likelihood) of things happening.

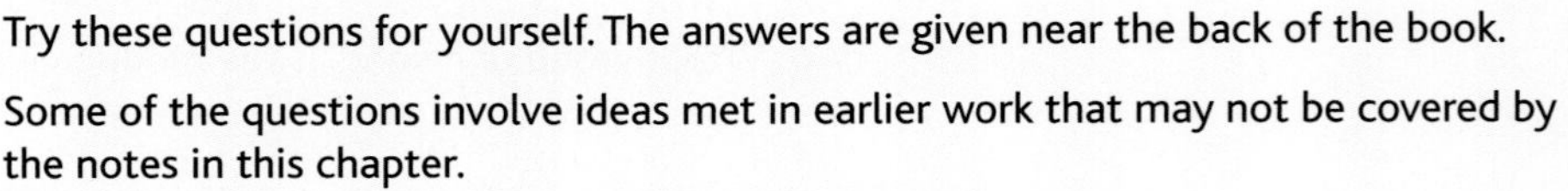

Exam-style questions

Try these questions for yourself. The answers are given near the back of the book.

Some of the questions involve ideas met in earlier work that may not be covered by the notes in this chapter.

11.8 Jamie has an ordinary coin and a triangular spinner.

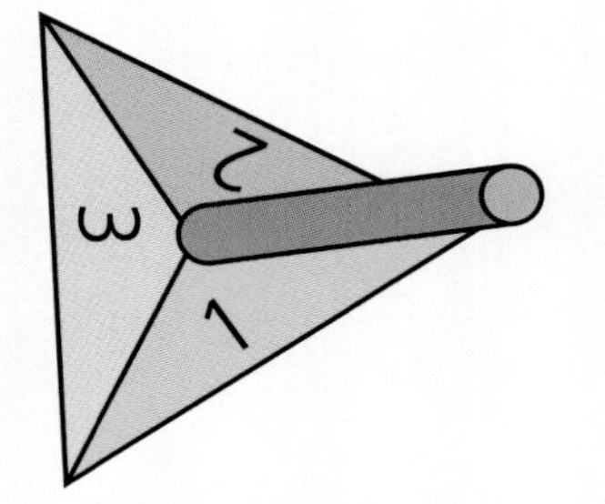

(a) How many possible outcomes are there when Jamie

(i) tosses the coin (1)

(ii) spins the spinner? (1)

(b) Jamie tosses the coin and spins the spinner at the same time.

One possible outcome is H (heads) and 2

List all the possible outcomes. (3)

11.9 Sally has two ordinary dice; one red and one blue.

(a) Describe all the ways it is possible for her to score a total of

(i) 5 (2)

(ii) 8 (2)

Use these tables to help you.

Score 5	
Red	Blue
4	1

Score 8	
Red	Blue

Sally and Minnie play a game with the two dice. To start, the player must roll two sixes. Minnie says it would be easier to roll two ones.

(b) Do you think Minnie is right? Explain your answer. (1)

11.10 Tina is using this spinner in a game.

Choose your answers from these:

impossible unlikely even chance likely certain

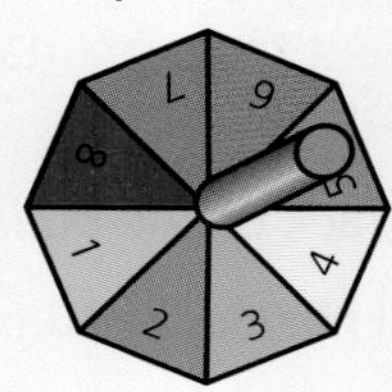

Which of the words best describes the probability of scoring:

(a) an even number (1)

(b) a number larger than 8 (1)

(c) 7 (1)

(d) a factor of 24 (1)

(e) a prime number? (2)

11.11 Mark with a cross on the probability scales the probabilities of these events.

(a) When a fair die is rolled, the score will be 5 (1)

impossible certain

(b) When a fair coin is tossed, it will land showing tails. (1)

impossible certain

(c) There will be 31 days in May next year. (1)

impossible certain

★ Make sure you know

- ★ How to collect **discrete data** and record them using a **tally** and a **frequency table**
- ★ How to use the **mode** and **range** to describe sets of data
- ★ How to group data, where appropriate, in equal class intervals, represent collected data in **frequency diagrams** and interpret such diagrams
- ★ How to use the **mean** of discrete data
- ★ How to use the **median** of a set of data
- ★ How to use your knowledge of **averages** to compare two **distributions**
- ★ How to construct and interpret simple **line graphs** and be able to read values from them
- ★ How to use simple vocabulary associated with probability, including **fair**, **certain** and **likely**
- ★ How to use the glossary at the back of the book for definitions of key words

Remember that you should be familiar with all of the material from the earlier years of the National Curriculum, for example, Venn diagrams and Carroll diagrams.

Test yourself ✓

Before moving on to the next chapter, make sure you can answer the following questions.

The answers are near the back of the book.

1 The scores here were achieved by 20 pupils in a mathematics quiz.

11	14	8	17	14	17	13	9	20	10
12	10	16	15	19	12	16	17	7	13

(a) For this set of marks, what is

(i) the range

(ii) the mode

(iii) the median?

The pupils were awarded grades as indicated in this table.

Grade	Scores	Tally	Frequency
A	17–20		
B	13–16		
C	9–12		
D	5–8		
E	1–4		

(b) Complete the table.

(c) Draw a frequency diagram to show the numbers of pupils awarded each grade.

(d) What is the mean of the top five scores?

2 The line graph below shows Paula's cycle ride to visit Jayne.

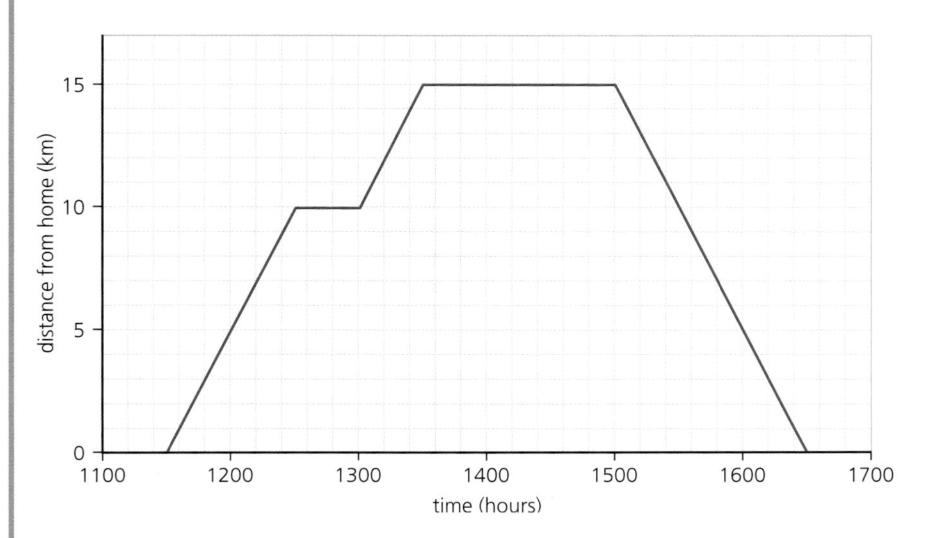

(a) At what time did Paula leave home?

Paula had a puncture and spent half an hour repairing it.

(b) How far from her home did Paula have the puncture?

(c) At what time did Paula reach Jayne's house?

(d) How long did Paula stay at Jayne's house before heading home?

(e) How far did Paula cycle altogether?

(f) For how long was Paula away from her home?

3 Jasmine has drawn this conversion graph to help her compare distances in miles and kilometres.

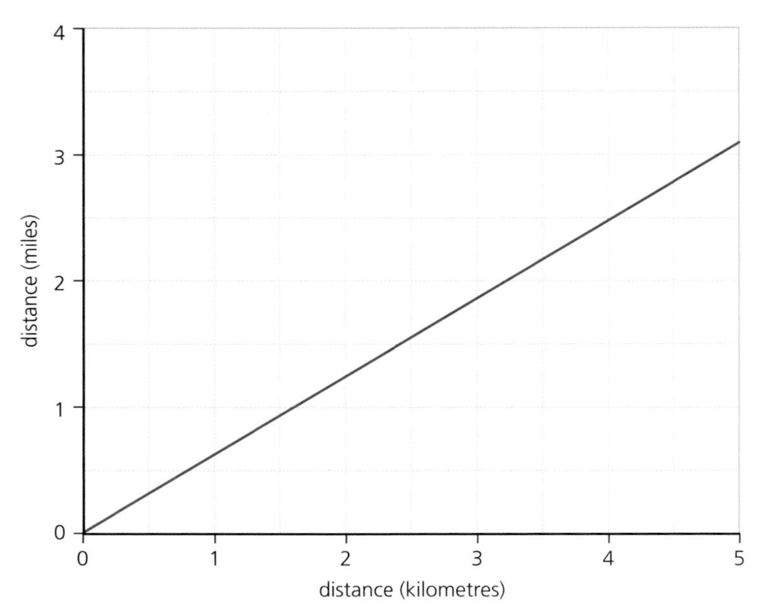

(a) Peter lives 4 km away from Jasmine. What is this distance in miles?

(b) Jasmine sets off to run a mile. What is this distance in metres?

4 Give an example of an event that

(a) is impossible

(b) is *very* unlikely

(c) is unlikely

(d) has an even chance of happening

(e) is likely

(f) is *very* likely

(g) is certain to happen.

5 At a funfair there was a stall where people could roll a die.

ROLL THE DIE
£1 to enter
Choose your own die
Roll a 6 and win £6

Tim thought about this and decided to have six tries, knowing that he would spend £6, and thinking that he would be bound to win £6 on one of his six rolls, so he could not lose! Since blue is his favourite colour, he chose a blue die and rolled it six times.

He was dismayed to discover that he did not roll a single six. The numbers he rolled were: 5, 2, 3, 1, 2, 4

Tim claimed that the die must be biased!

(a) (i) What do you think about Tim's strategy?

(ii) Do you think the die was biased?

The stallholder told Tim to choose another die and try again, free of charge. This time Tim chose a large gold coloured die and rolled 1 six times!

(b) Do you think that this second die is fair?

(c) Describe a sensible way of finding out if a die is biased.

(d) Suggest a way in which the stallholder can be fairly certain that he will make money.

Statistics and probability (2)

12.1 Statistics

Drawing conclusions from graphs, charts and diagrams

Whatever the type of graph, it is important to look at the whole picture before you try to conclude anything. Many graphs are deliberately misleading because they have something to hide – the newspapers are full of examples – do not be deceived!

Example

Which country below has more rain?

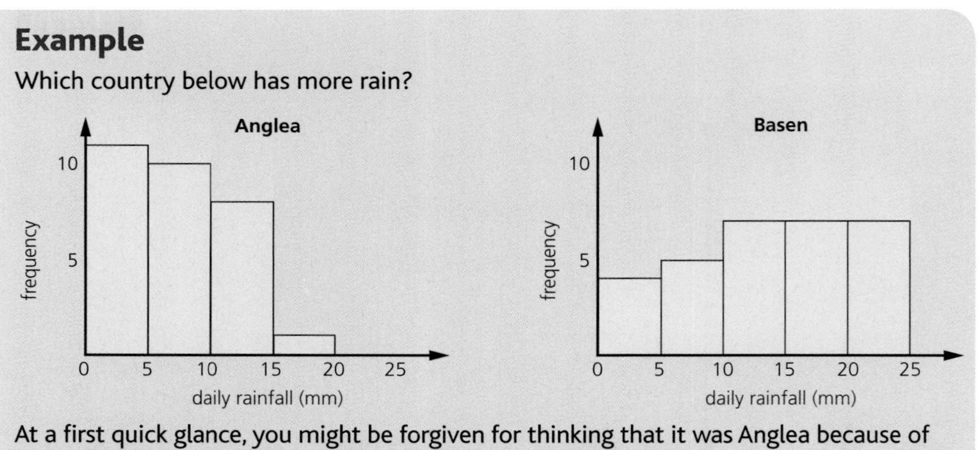

At a first quick glance, you might be forgiven for thinking that it was Anglea because of its tall bars. However, it is in fact Basen, because it has fewer days of low rain (0–5 mm) and more days of high rain (20–25 mm) than Anglea.

Frequency diagrams (bar charts) using discrete data

Example

The first 20 draws of the UK National Lottery produced these numbers (7 each week) in the order given. Bonus numbers are printed in bold type.

30	3	5	44	14	22	**10**	16	6	44	31	12
15	**37**	21	11	17	30	29	40	**31**	26	47	49
43	35	38	**28**	13	3	38	5	14	9	**30**	27
29	39	3	44	2	**6**	17	44	36	32	9	42
16	21	32	2	5	25	22	**46**	23	38	17	7
32	42	**48**	47	6	16	31	30	20	**4**	31	16
25	43	4	26	**21**	46	42	1	38	7	37	**20**
48	38	15	29	18	35	**5**	45	16	36	19	21
29	**43**	18	33	8	31	5	10	**28**	17	36	11
12	42	26	**13**	2	22	13	46	29	27	**36**	41
19	31	18	9	24	**21**	4	49	41	44	42	17
24	43	41	22	25	30	32	**29**				

To make a frequency chart of these numbers, we first draw up a frequency table.

Note that we have had to make the first group (1 to 9), only 9 numbers instead of 10, because there is no zero ball.

Numbers drawn	Frequency
1 to 9	24
10 to 19	28
20 to 29	30
30 to 39	30
40 to 49	28

Well, it seems fair. That first bar is a bit shorter than the others because, of course, it is counting only 9 balls not 10, as mentioned above.

Mathematical pause for thought: every time somebody becomes a lottery millionaire, it means that between them the others have lost over a million pounds.

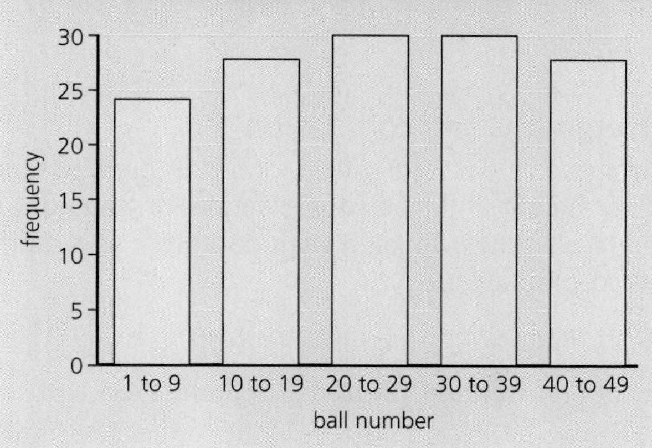

Pie charts

Example

Twenty children were asked about their favourite ice cream. The replies were as follows:

chocolate: 9; vanilla: 6; strawberry: 4; toffee: 1; total: 20

Display the results of the ice cream survey using a pie chart.

For a pie chart measured in degrees, this total of 20 must be made into 360°. For a pie chart measured in percentages, it must be made into 100%. We shall work through both calculations here.

Degrees version

360° ÷ 20 = 18°, so each child is represented by 18°

Flavour	**Frequency**	**Factor**		**Angle of sector**
chocolate	9	× 18°	=	162°
vanilla	6	× 18°	=	108°
strawberry	4	× 18°	=	72°
toffee	1	× 18°	=	18°
total	20		**total**	360°

Percentage version

100% ÷ 20 = 5%, so each child is represented by 5%.

Flavour	**Frequency**	**Factor**		**Size of sector**
chocolate	9	× 5%	=	45%
vanilla	6	× 5%	=	30%
strawberry	4	× 5%	=	20%
toffee	1	× 5%	=	5%
total	20		**total**	100%

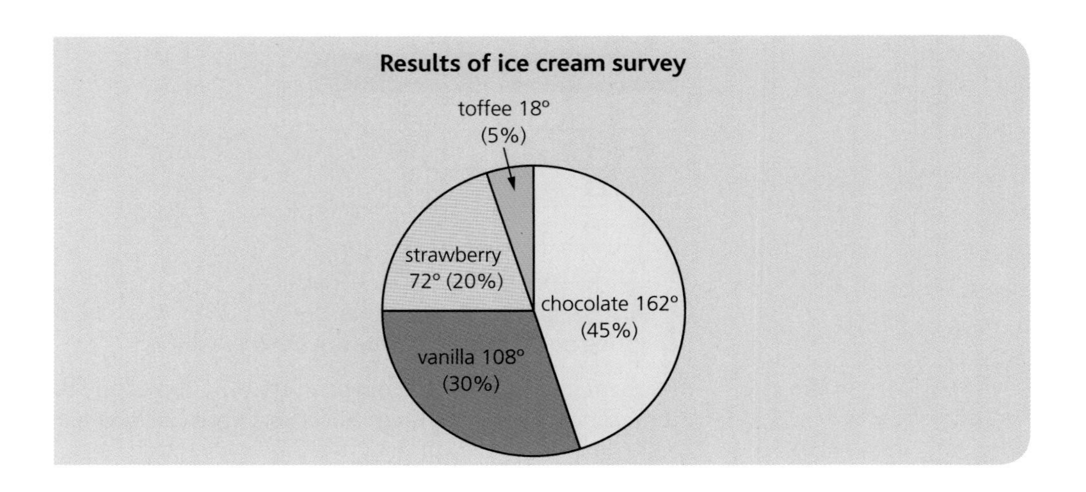

Scattergraphs and correlation

Sometimes the data we collect comes in pairs of numbers. We might ask people for their height and mass, for example, or we might measure oxygen levels at different altitudes. To keep each data item as a pair, we plot them as Cartesian (x, y) co-ordinates:

(x, y) representing (height, mass) or

(x, y) representing (altitude, oxygen level) and so on.

When we have plotted several points, we end up with a graph called a **scattergraph**.

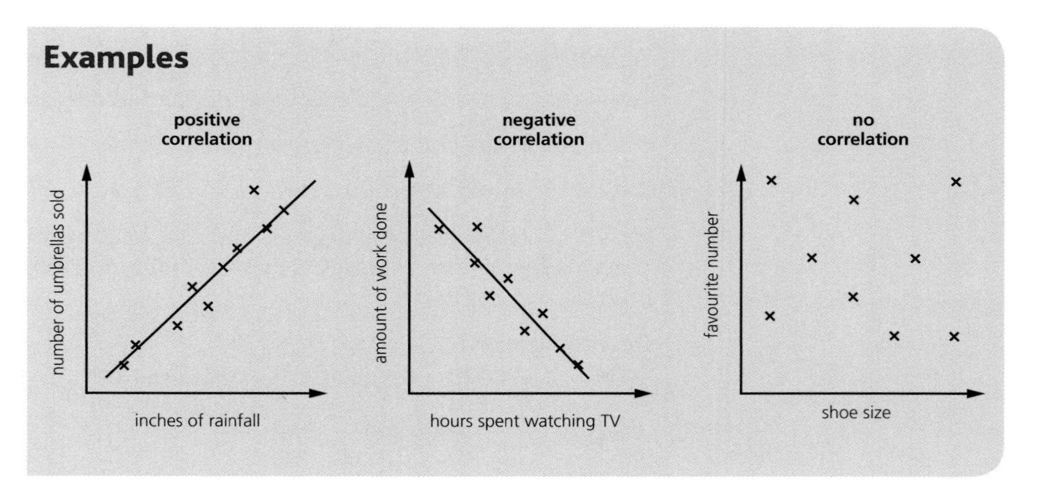

Note that when there is either positive or negative **correlation**, we can draw a line showing the approximate direction of the trend in the points.

This line, drawn by eye, is called the **line of best fit**.

When there is correlation (positive or negative), it means that the pairs of numbers are related in some way.

Sets and Venn diagrams

Sets

Sets are simply collections of things, all of which have something in common. For example, we might talk about the set of odd numbers less than 20, or the set of prime numbers between 10 and 30

With Venn diagrams we can show things that are in more than one set at the same time.

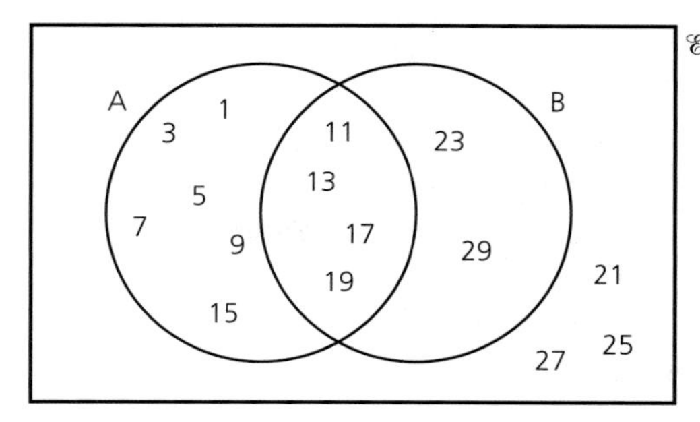

$\mathscr{E}$ = {odd numbers less than 30}

A = {odd numbers less than 20}

B = {prime numbers between 10 and 30}

Set notation is unlikely to be included in the ISEB Common Entrance exam.

Set notation

A list of signs and symbols used in describing sets:

$\in$ is a member (element) of

$\notin$ is not a member of

$\cup$ union: elements in either one set or the other (or both) are included

$\cap$ intersection: only elements in *both* sets are included

$\subset$ is a subset of

$\supset$ contains

$\varnothing$ the empty set (contains no elements)

A′ complement: those items *not* in set A

$\mathscr{E}$ the universal set: this set contains all elements under consideration

We list the elements in a set within curly brackets, e.g. A = {a, e, i, o, u}.

We count the number of elements in a set using $n()$, as in $n(\text{A}) = 5$ for the set A above.

Sets and Venn diagrams are unlikely to be included in the ISEB Common Entrance exam.

Sets and Venn diagrams

These Venn diagrams show how the signs and symbols listed above can be represented visually. In the first six diagrams, the notation below the diagram describes the shaded area.

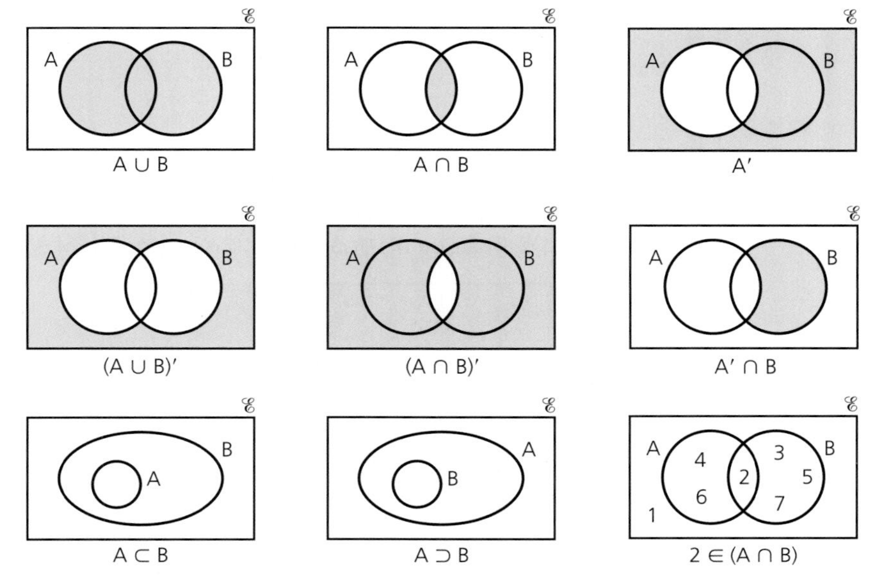

It is useful to be able to answer questions using set notation.

Example

Look again at the Venn diagram showing the odd numbers less than 30

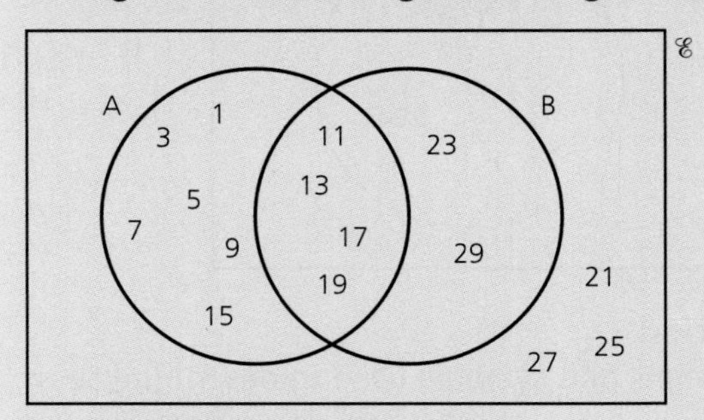

(i) List the members of the set $A \cap B$.

(ii) Find $n(A')$.

(iii) How would you describe the numbers in the set $(A \cup B)'$?

Answers

(i) 11, 13, 17, 19

(ii) 5

(iii) Numbers between 20 and 30 which are odd but not prime.

Exam-style questions

Try these questions for yourself. The answers are given at the back of the book.

Some of the questions involve ideas met in earlier work that may not be covered by the notes in this chapter.

12.1 The 38 boys in Year 8 were asked which sports teams they played in. The results are shown in this Venn diagram.

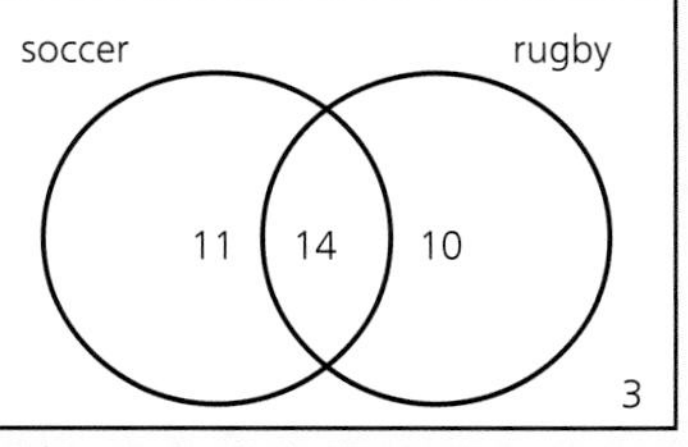

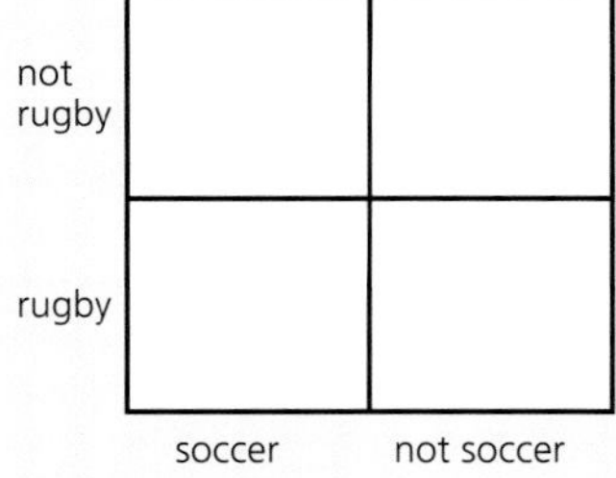

(a) Record the same results on the Carroll diagram here. (2)

(b) How many boys in Year 8 play in just one team? (1)

12.2 This fraction diagram shows the proportion of 3 metals in an alloy.

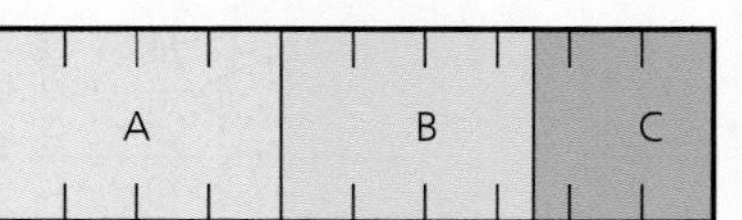

(a) What fraction, in its simplest form, of the alloy is metal A? (2)

(b) What percentage of the alloy is metal B? (2)

(c) If a pie chart was drawn to show the same information, what angle would represent metal C? (1)

12.3 This pie chart represents the favourite ice cream flavours chosen by 30 people.

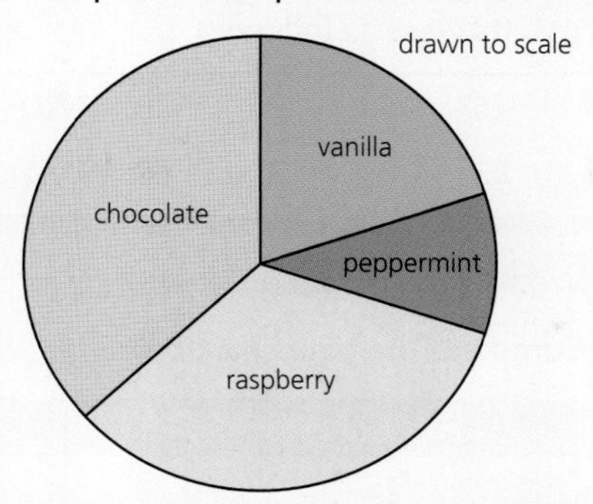

(a) How many degrees represent each person? (1)

(b) How many people chose peppermint? (1)

(c) How many more people chose chocolate than chose raspberry? (2)

(d) What percentage of the people chose vanilla? (2)

12.4 This frequency diagram shows the scores in a school quiz.

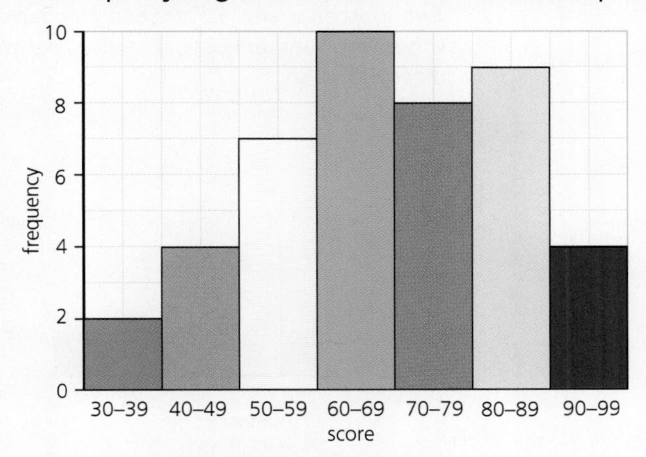

(a) Which **class interval** is the **mode**? (1)

(b) How many pupils scored 70 or more? (1)

12.5 Sandy has rolled an ordinary die 20 times and recorded his scores.

3	5	6	1	6	1	4	1	3	6
5	2	6	4	2	4	5	2	6	2

(a) Fill in the tally chart. (3)

Score	Tally	Frequency
1		
2		
3		
4		
5		
6		
	Total	

(b) Draw a frequency diagram to show this information. (3)

(c) What is Sandy's modal score? (1)

(d) What is his mean score? (3)

(e) What is his median score? (2)

12.6 Each girl in a class of 18 girls counted how many peas she had on her plate. The results were as follows:

27	25	30	26	28	20	24	26	25
25	22	27	25	29	31	26	28	21

(a) What is the modal number of peas? (1)

(b) Complete the tally chart below. (3)

Number of peas	Tally	Frequency
20–22		
23–25		
26–28		
29–31		
	Total	

(c) What is the modal class? (1)

12.7 Two football teams, Rovers and Wanderers, have played each other 12 times. The numbers of goals scored per match are shown in these frequency diagrams.

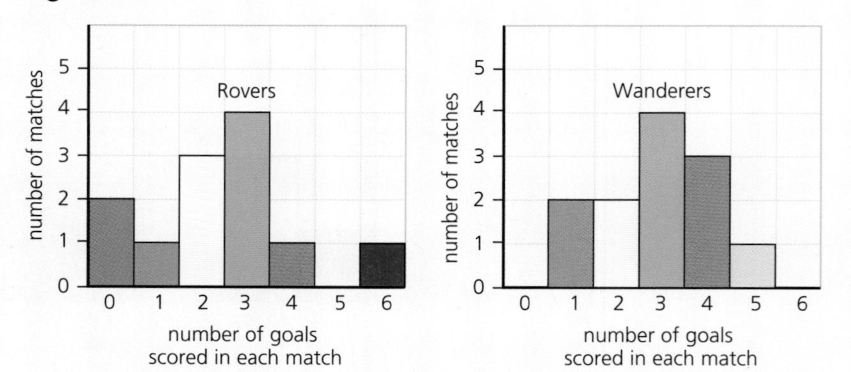

(a) **(i)** What is the range of the number of goals scored per match by Rovers? (1)

(ii) What is the modal number of goals scored per match by Wanderers? (1)

(iii) What is the median number of goals scored per match by Rovers? (2)

(b) Compare the performance of the two teams, making use of the frequency diagrams and any additional calculations you have made. (2)

12.8 The mean mass of ten children is 36.2 kg.

(a) What is the total mass of the children? (2)

When two of the children leave, the mean mass of the eight remaining children falls to 35 kg.

(b) What is the total mass of the eight remaining children? (1)

(c) What is the mean mass of the two children who left the class? (2)

12.9 Some children were asked to state their favourite breakfast cereal.

The results, recorded here, are to be represented on a pie chart.

puffs	30%
crispies	15%
flakes	20%
muesli	the rest

(a) Draw a pie chart, naming each sector clearly and giving the angles. (4)

(b) If 60 children took part in the survey:

(i) How many children chose crispies as their favourite cereal? (2)

(ii) How many children chose muesli as their favourite cereal? (2)

12.10 Below is a graph that could be used for converting British pounds (£) into US dollars ($) one day in January 2015

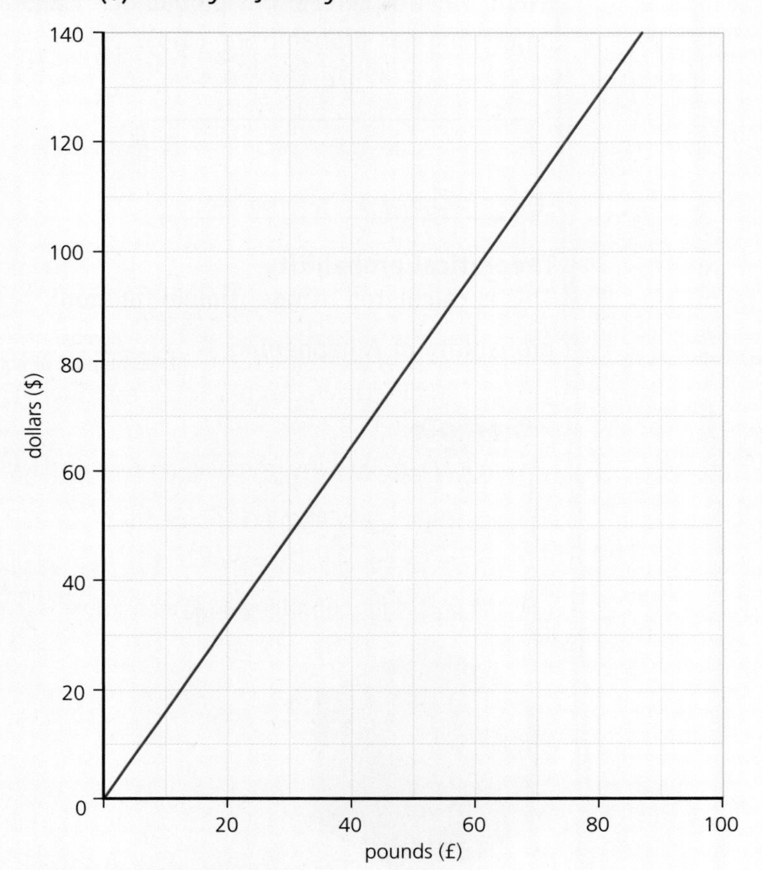

(a) Use the graph to find:

(i) how many dollars were equivalent to £35 (2)

(ii) how many pounds were equivalent to $80 (2)

(b) On a flight to New York, a duty free watch was priced at £65

In the airport duty free shop, the same item was priced at $80

(i) Which price was the better value? (2)

(ii) What was the difference in the prices? Give your answer in GBP to the nearest 1 pound. (2)

12.11 The masses and heights of ten girls are recorded in the table below.

Mass (kg)	30	40	50	42	56	38	35	36	39	33
Height (cm)	165	140	145	148	144	135	142	150	152	148

(a) Plot points representing this information as a scattergraph. (3)

(b) Draw, *if appropriate*, the line of best fit. If this is not possible, then explain why not. (1)

(c) Using your line of best fit, or otherwise, estimate the height of a girl of mass 45 kg. (2)

12.2 Probability

Theoretical and experimental probability

Likelihood and probability

We can add numbers to the likelihood scale (Chapter 11, Section 11.2, Words to describe probability), so that an event which is impossible has probability 0, an event which is certain has probability 1 and an event with an even chance of happening has probability $\frac{1}{2}$ (0.5).

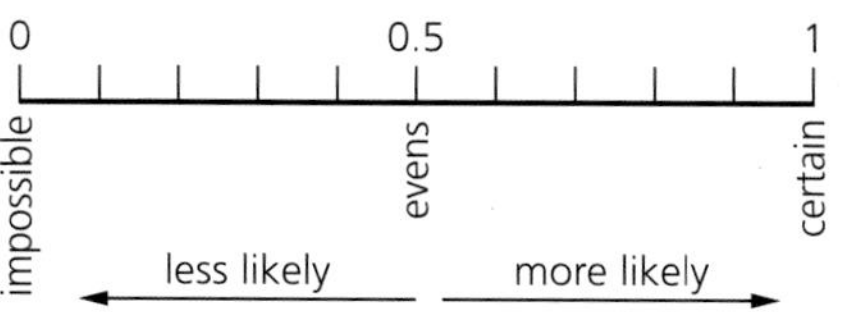

Theoretical probability

This is calculated using a simple fraction:

$$\text{probability of A happening} = \frac{\text{number of ways A can happen}}{\text{total number of equally likely possibilities}}$$

Example

What is the probability of rolling an even number with an ordinary die with six faces?

It can happen in three ways (2, 4 or 6).

There are six equally likely possibilities in total (1, 2, 3, 4, 5 and 6).

So the probability of rolling an even number is $\frac{3}{6}$, or 0.5 (50%).

Experimental probability

This is calculated using a similar fraction:

$$\text{probability of X happening} = \frac{\text{number of times X happened}}{\text{total number of experiments}}$$

Example

The drawing-pin experiment is a familiar example. Will the pin land point up or on one side? Try it and see!

Probability calculations

The sum of all probabilities in a situation is 1

We saw earlier that certainty had a probability of 1

When we say that the sum of all probabilities is 1, all we are saying in effect is that it is certain that *something* will happen. Quite obvious really!

Let's see now how useful that fact can be.

Example 1

The probability that Tom will be picked for the school hockey team is 0.7

What is the probability that he won't be picked?

The total probability (will + won't) is 1

The probability that he will be picked is 0.7

So the probability that he won't is 1 – 0.7 = 0.3

Example 2

When I phone my friend Anne, the probability that she will answer is 0.3 and the probability that someone else in her family will answer is 0.4

The probability that no one answers is 0.1; the only other possibility is that the line is engaged. What is the probability that the line is engaged?

The sum of all these probabilities is 0.8

So the probability that the line is engaged is 0.2

Listing outcomes systematically

There are various ways of listing outcomes:

Method 1

How many different four-digit numbers can you make using 7, 3, 1 and 4 once only?

Well, there's 7314 and 3741 and 1743 and ...

It would be difficult to carry on like this without missing numbers out or repeating earlier ones. Being systematic means being organised in your approach. Like this, for example:

1347	1374	1437	1473	1734	1743
3147	3174	3417	3471	3714	3741
4137	4173	4317	4371	4713	4731
7134	7143	7314	7341	7413	7431

Read across the rows above. Can you see the method?

Method 2

The table below shows all the possible outcomes when two dice are rolled and the scores added together. It then becomes very easy to see that a total of 7 can be reached in six different ways, making it the most likely total, but 2 and 12 can only be made in one way, making them the two least likely totals.

+	1	2	3	4	5	6
1	2	3	4	5	6	7
2	3	4	5	6	7	8
3	4	5	6	7	8	9
4	5	6	7	8	9	10
5	6	7	8	9	10	11
6	7	8	9	10	11	12

Method 3

What are the possible outcomes when three coins are tossed onto a desk at the same time?

Possibility	1	2	3	4	5	6	7	8
Coin 1	H	H	H	H	T	T	T	T
Coin 2	H	H	T	T	H	H	T	T
Coin 3	H	T	H	T	H	T	H	T

Once again, careful listing shows all eight possibilities very clearly. What patterns can you find in the way the letters H and T have been written in the table?

Exam-style questions

Try these questions for yourself. The answers are given at the back of the book.

Some of the questions involve ideas met in earlier work that may not be covered by the notes in this chapter.

12.12 Samantha has a bag containing 10 sweets: 4 red, 1 yellow and 5 blue.

She picks one sweet at random.

(a) (i) What is the probability that Samantha picks a red sweet? (1)

(ii) Record this probability, with a letter R, on the probability scale below. (1)

0 |—|—|—|—|—|—|—|—|—|—| 1

The sweet is red and Samantha eats it. She then picks another sweet.

(b) What is the probability that Samantha picks a red sweet this time? (1)

12.13 The results of spinning a hexagonal spinner 100 times are recorded in this frequency table.

Score on spinner	Frequency
1	18
2	13
3	10
4	13
5	17
6	29

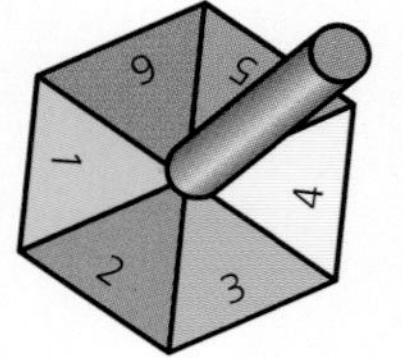

(a) Do you think the spinner is **biased**? Explain your answer. (1)

(b) Based upon the experimental results, suggest, as a percentage, the probability that this spinner will score 6 (1)

12.14 Stephen rolls two tetrahedral dice, one red and one green, each with faces numbered 1, 2, 3 and 4

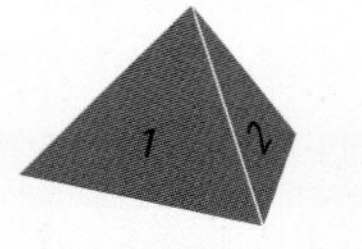

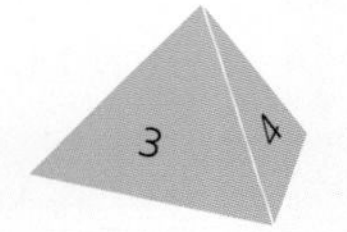

He adds the scores together.

(a) Complete the table below to show all possible total scores. (2)

		Score on red die			
		1	2	3	4
Score on green die	1	2			
	2			5	
	3				
	4		6		

(b) What is the probability of Stephen's total score being:

(i) 4 (1)

(ii) more than 3? (1)

Stephen rolls the red die and scores 3

(c) What is the probability that his score on the green die will give him a total score of 5 or more? (2)

12.15 Marie and Clare are playing a game. Each girl has written the letters of her name on separate cards and placed the cards in her bag.

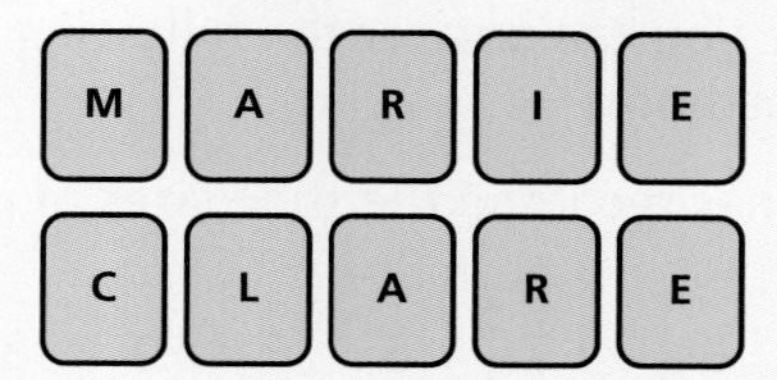

Each girl takes a card at random from her bag. Marie takes R and Clare takes L.

They replace the cards and repeat the process four times. The results are recorded in the table below.

	C	L	A	R	E
M	MC				
A			AA		
R		RL			
I					IE
E				ER	

(a) Complete the table above to show all the possible combinations. (3)

(b) What is the probability that the two cards taken will have the same letter? (2)

(c) What is the probability that the two cards taken will both have a vowel? (2)

★ Make sure you know

- ★ How to compare two simple distributions, using the range and one of mode, median or mean
- ★ How to interpret graphs and diagrams, including pie charts, and draw conclusions
- ★ How to construct and interpret frequency diagrams
- ★ How to construct pie charts, using degrees or percentages
- ★ How to draw conclusions from scatter diagrams, and have a basic understanding of correlation
- ★ How to draw a line of best fit on a scatter diagram, by inspection
- ★ How to use the probability scale from 0 to 1
- ★ The difference between experimental and theoretical probability
- ★ How to find and justify probabilities, and approximations to these, by selecting and using methods based on equally likely outcomes and experimental evidence, as appropriate
- ★ That different outcomes may result from repeating an experiment

★ How to identify all the outcomes, using diagrammatic, tabular or other forms of communication, when dealing with a combination of two experiments

★ How to use your knowledge that the total probability of all the mutually exclusive outcomes of an experiment is 1

★ How to use the glossary at the back of the book for definitions of key words

Test yourself

Make sure you can answer the following questions. The answers are at the back of the book.

1 Gerry has recorded the shoe sizes of nine friends.

$6\frac{1}{2}$ $8\frac{1}{2}$ $7\frac{1}{2}$ 6 7 9 $8\frac{1}{2}$ $9\frac{1}{2}$ 8

(a) What is the range of these shoe sizes?

(b) Find the median shoe size of Gerry's friends.

(c) What is the modal shoe size?

2 The mathematics examination marks for a class are shown below.

64 79 53 45 63 78 66 77 47 67
73 69 90 58 51 42 75 69 67 63

(a) What is the range of marks for this set of results?

Mr Hillard decides to sort the marks into groups: 40–49, 50–59 and so on.

(b) Complete the tally and frequency columns in the table below.

Marks	Tally	Frequency
40–49		
50–59		
60–69		
70–79		
80–89		
90–99		
	Total	

(c) What was the modal class of marks in the examination?

(d) Draw a frequency diagram to show the results.

Here is a frequency diagram showing the marks for the same class in an earlier mathematics examination.

(e) By comparing the two graphs, comment on the performance of the class in the two examinations.

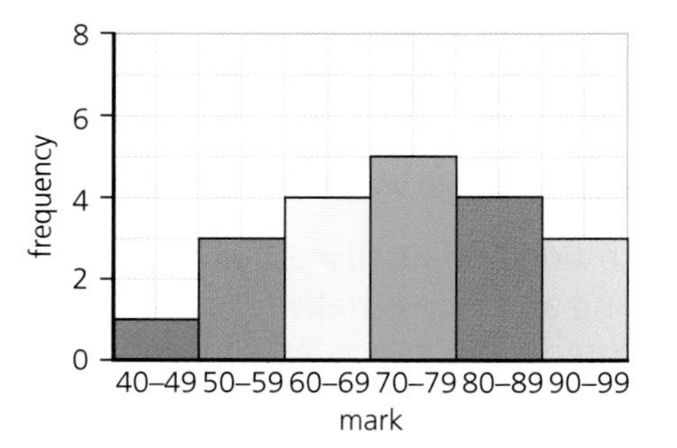

3 A garden centre sells boxes of mixed bulbs. Each box of bulbs contains a gross (144) of which $\frac{1}{4}$ are daffodils and $\frac{1}{16}$ are hyacinths. In addition, there are three dozen snowdrops, twice as many tulips as hyacinths and the rest are crocuses.

(a) Draw a fully-labelled pie chart to show this information, marking each sector clearly with both angle and name of bulb.

A different garden centre also sells boxes containing a gross of mixed bulbs.

On a pie chart showing the selection of bulbs in these boxes, daffodils are represented by a sector of angle 135°.

(b) What percentage of the bulbs in this box are daffodils?

4 The conversion graph below can be used to change centimetres to inches.

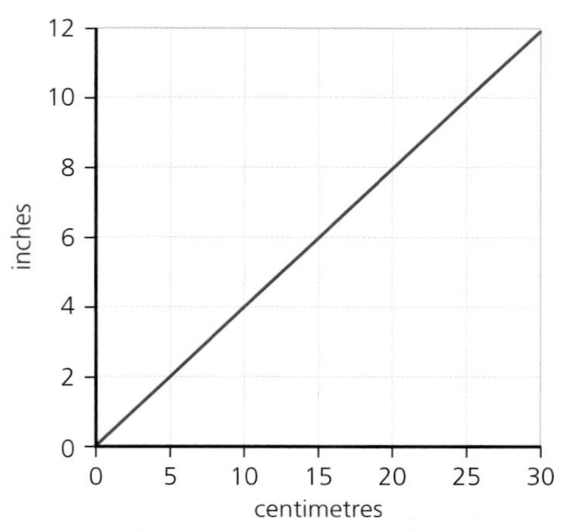

Use the graph to answer the following questions.

(a) How many inches are equivalent to 260 centimetres?

(b) There are 12 inches in 1 foot. Simon's height is 5 feet 10 inches. What is his height measured in centimetres?

5 The science and mathematics percentages of nine pupils are recorded in the table below.

Science mark	56	48	50	65	54	78	91	55	87
Mathematics mark	63	55	58	64	49	72	80	64	92

(a) Plot points representing this information as a scattergraph.

(b) Draw, by eye, the line of best fit.

(c) Using your line of best fit, estimate the mathematics mark for a pupil who scored 60 in science.

6 In the National Lottery, the machine can choose at random any one of the first 49 counting numbers (1 to 49 inclusive) when selecting the first number.

(a) What is the probability that the first number selected will be:

(i) the number 23

(ii) an even number

(iii) a prime number less than 20?

The first number chosen is 13 and the machine now chooses at random a second number.

(b) What is the probability that the second number chosen will be 30?

7 On a snooker table there are 15 red balls and one each of white, yellow, green, brown, blue, pink and black.

Hal, who is blindfolded, picks one ball at random from the table.

(a) What is the probability that the ball is

(i) yellow

(ii) red

(iii) not blue?

Hal picked the green ball and put it in his pocket.

Jenny now removes a third of the red balls, the blue ball and the pink ball.

Hal then picks another ball at random.

(b) What is the probability that he picks a red ball?

8 In the Venn diagram shown here:

$\mathscr{E}$ = {integers from 40 to 50 inclusive}

A = {odd numbers}

B = {multiples of 5}

C = {multiples of 7}

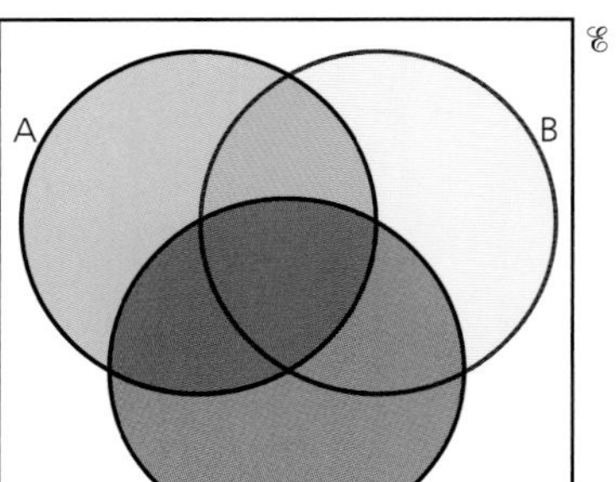

(a) List all the members of

(i) set A

(ii) set B

(iii) set C.

(b) Complete the Venn diagram.

(c) A number is chosen at random from $\mathscr{E}$. What is the probability that the number chosen is a multiple of both 5 and 7?

Glossary

Abacus	The predecessor of the electronic calculator: a device usually consisting of rows of beads on wires to assist with mental calculations.
Acre	A measurement of area in the imperial system of units (about 4000 m^2).
Acute angle	An angle less than 90 degrees.
Add	Plus; find the total.
Addition	The process of adding; a 'sum'.
Adjacent	Next to; side by side; in a right-angled triangle the side which is not 'opposite' (to a given angle).
Algebra	The branch of mathematics using letters to stand for numbers.
Algebraic expression	One or more algebraic terms combined using + and −, e.g. $5 + 2 - x$
Alternate angles	Equal angles formed where one line crosses a pair of parallel lines, creating a 'Z' or 'N' figure.
a.m.	*ante meridiem* (Latin) which means 'before noon'; see also **p.m.***
Amount	Quantity; how much or how many.
Angle	The amount in degrees by which two joined lines have been 'hinged open'.
Angle measurer	An instrument such as a protractor for measuring the sizes of angles.
Angles at a point	Two or more angles sharing a common vertex with an angle sum of 360°. $a + b + c = 360°$
Angles in a triangle	The three angles in the same triangle always add up to 180°; in a right-angled triangle this is equivalent to saying that the other two angles always add up to 90°. $a + b + c = 180°$ $d + e = 90°$
Angles on a straight line	Two or more angles sharing a common edge with an angle sum of 180°. $f + g = 180°$
Answer	That which was sought by the question.
Anticlockwise	An opposite direction to that in which the hands of a clock turn.
Apex	The highest point in a plane or solid shape.
Approximately	By way of an approximation; roughly.
Approximately equal to (≈)	Roughly the same as; having used approximate rather than exact values.
Approximation	A 'rough' answer, often used as a check to a calculator result.

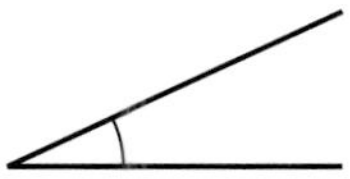

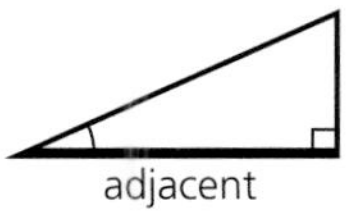

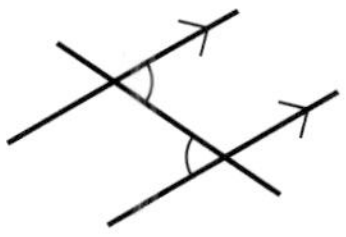

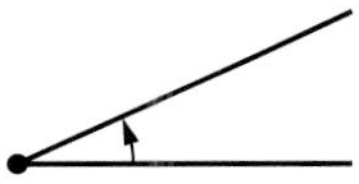

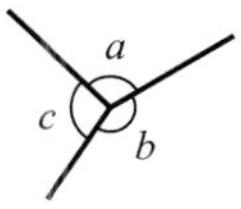

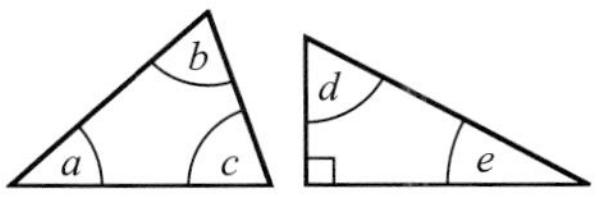

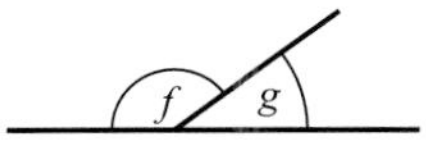

adjacent

$a + b + c = 360°$

$f + g = 180°$

*Bold type means that a word has its own glossary entry as well as appearing in this definition.

Arc A section of the circumference of a circle.

Area The amount of space inside a flat shape, commonly measured in square centimetres (cm^2), square metres (m^2), hectares (ha), etc.

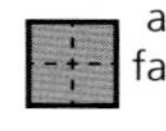

Area scale factor The old area was multiplied by this to obtain the new area; the area scale factor is the square of the scale factor.

Arithmetic The branch of mathematics dealing with numbers, especially with the operations +, −, ×, ÷

Arithmetic sequence A sequence of numbers which goes up (or down) by the same amount each time, e.g. 1, 4, 7, 10, …

Arrangement A way of putting things in an order or an 'array' – a pattern.

Arrowhead The only concave quadrilateral (a type of kite); delta or nested V-shape.

Ascending Going up; in order of increasing size.

Average A calculation used to summarise a set of data.

Average speed The total distance travelled divided by the total time taken.

Axes Plural form of **axis**.

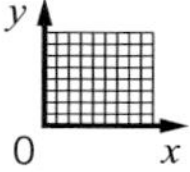

Axis A line on a grid holding the numbers of scale, usually horizontal (x-axis) and vertical (y-axis).

Axis of rotation symmetry Imagine a CD spinning around a pencil – the pencil is the axis of rotation symmetry.

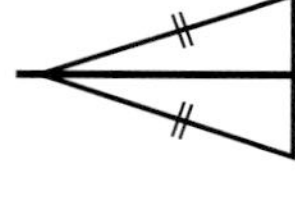

Axis of symmetry Also known as the line of symmetry.

Balance (1) The remainder after a series of financial payments (in and out).

Balance (2) An old-fashioned pair of weighing scales: if the scales balanced then the masses were equal.

Bank balance The amount of money in the bank.

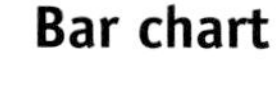

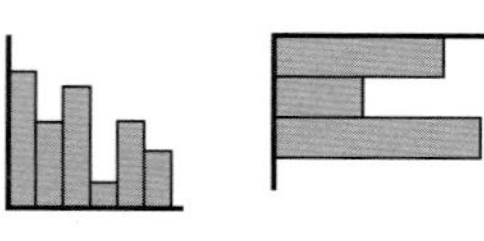

Bar chart A diagram which shows data by the height or length of the bar.

Bar-line graph As a bar chart, but using lines or sticks, usually to represent frequency (also called a 'stick graph').

Base The bottom face of a solid shape or the lower edge of a plane shape.

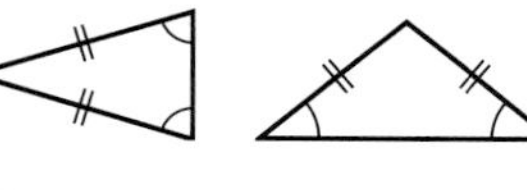

Base angles Especially in an isosceles triangle – the two angles which are equal. (NB base angles are not necessarily at the bottom of the picture!)

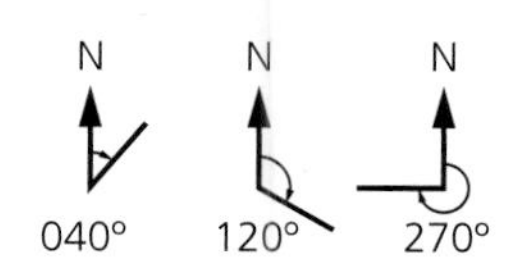

Bearings A three-figure direction measured clockwise in degrees from north.

Best estimate The answer which is closest to the exact answer without having to work it out.

Between Somewhere along the line from one to the other.

Bias A tendency to go one way more than another; favouritism.

Biased A die, spinner or coin is biased if it is not fair.

BIDMAS see **BODMAS**.

Billion One thousand million, written as 1 000 000 000

Bisect Cut into two equal halves; the two diagonals of a rectangle, square, rhombus and parallelogram bisect each other.

Bisector A line which cuts something in half, commonly an angle or another line.

BODMAS Brackets, Of, Division, Multiplication, Addition, Subtraction – a way to remember the priorities of arithmetical operations. (Some people prefer **BIDMAS**, where the second letter is for Indices.)

Brackets Another name for parentheses '(' and ')'; see also **Factorise** and **Multiply out**.

Calculate Work out; evaluate. (It does not always mean use a calculator!)

Calculation The process of finding the value of a numerical expression.

Calculator Any machine which makes calculations easier.

Cancel Cross out, usually in pairs, and especially when simplifying fractions or ratios.

e.g. $\frac{\cancel{4}^{2}}{\cancel{6}_{3}} \rightarrow \frac{2}{3}$ $\quad {}^{3}\cancel{6} : \cancel{2}^{1} \rightarrow 3 : 1$

Cancellation The process of simplifying which involves dividing by a common factor.

$\overset{1}{\cancel{6}}x = \overset{2}{\cancel{12}}$

$\rightarrow x = 2$

Capacity The amount of space taken up or enclosed by a solid; the volume.

Carroll diagram A two-way table used to help with classification, e.g. Girls (G) / Boys (B) against Swimmers (S) / Non-Swimmers (N). (Named after Lewis Carroll.)

Celsius The temperature scale, formerly known as centigrade, on which water freezes at 0 degrees and boils at 100 degrees.

Centilitre One hundredth of a litre.

Centimetre One hundredth part of a metre.

Centre The middle.

Centre of enlargement The point from which the enlargement is constructed; the 'focus' or 'vanishing point'.

Centre of rotation The part of the picture which does not move when the picture is rotated, e.g. the hub of a wheel.

Century A period of time lasting one hundred years.

Certain Having a probability of 1; impossible to fail; guaranteed to happen.

Chance Another word for probability or likelihood.

Change (money) The amount of money required to make up the cost of the purchases to the value of the cash offered.

Chart A picture which helps to represent a set of data.

Chord A straight line joining any two points on the circumference of a circle.

Circle The round shape made by joining up all points a fixed distance from the centre.

Circumference The perimeter of a circle.

Class interval The size of the group (when data is grouped), e.g. 20 to 25 cm has a class interval (width) of 5 cm.

Classify Sort out according to the properties of the items; books can be classified by genre, shapes can be classified by number of sides etc.

Clear (display, entry) Reset a calculator to receive the number again (C) or the whole calculation again (AC).

Clock (1) 12-hour clocks tell the time using only hours 1 to 12 but also with a.m. or p.m.

Clock (2) 24-hour clocks tell the time using hours 00:00 to 23:59; times after 12:00 are p.m. times.

Clockwise The same direction as that turned by the hands of a clock.

Coefficient The number in front of a letter or term, especially in an equation, e.g. $3x^2$ or $24x$

Collect like terms Group together terms in algebra with the same letter part, e.g. $3ab + 4a + 2ab = 4a + 5ab$

Column The vertical in a table (up – down).

Common factor A factor which is shared by two or more numbers or terms, e.g. 5 is a common factor of 15 and 95, and 4 is a common factor of $4a$ and $8b$

	Compare	Put one thing against another to see how similar or different they are.
	Compass	A direction finder marked as N, NE, E, SE, S, SW, W, NW.
	Compasses	Geometrical instrument for drawing circles; also used to draw arcs when constructing triangles.
	Complementary angles	Two angles which add up to 90 degrees. $a + b = 90°$
$a + b = 90°$	**Complements**	Fits the gap left by the other, especially in the context of complementary angles.
	Composite (1)	A shape made up from other simpler shapes.
	Composite (2)	A number which is not prime.
	Concave	Curving in (think 'caving in') like a letter C; the opposite to convex.
	Conclude	Sum up the main points of a reasoned argument or proof.
	Conclusion	The summary statement made from consideration of the statistics (figures and graphs) in the question.
	Cone	A circular-based pyramid.
	Congruence	The property of being congruent.
	Congruent	Identical in size and shape.
	Conjugate	Two angles which add up to 360 degrees; compare with complementary and supplementary.
$a + b = 360°$		
	Consecutive	Following on in counting order, e.g. 4, 5, 6 are consecutive numbers.
	Construct	Draw accurately a geometrical figure using a sharp pencil and geometrical instruments such as a ruler and compasses.
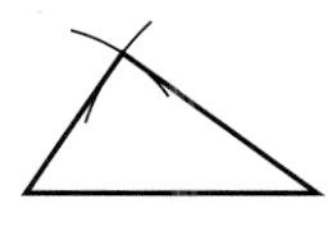	**Construction lines**	Lines left on a construction to show where the compasses and ruler were used.
	Continue	Carry on in the same way as before.
	Continuous	Without jumps or gaps; a measurement which allows 'in between' values.
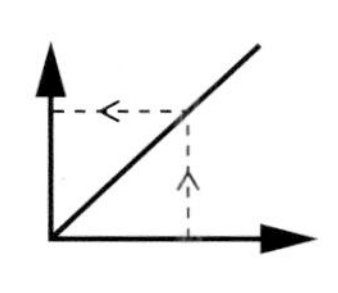	**Conversion graph**	A straight line graph used to convert between one unit and another, e.g. metric and imperial.
	Convert	Change from one into another, especially from one unit of measurement to another.

Convex Bulging outwards like a letter D; opposite to concave.

Co-ordinate pair Two numbers (x, y) used to plot a point on a co-ordinate grid.

Co-ordinate point The point plotted by a co-ordinate pair such as (2, 4).

Co-ordinates Pairs of numbers used to indicate positions on a grid; (x, y) = (across, up).

Correlation Describes the way one measurement increases relative to another; see also positive correlation and negative correlation.

Corresponding (points) If a shape has been transformed (e.g. by translation) then each vertex has a Start and Finish position; these two positions make a pair of corresponding points/vertices.

Corresponding angles In an angle diagram, the equal angles formed by a straight line crossing a pair of parallel lines ('F-angles').

Cosine (cos) In a right-angled triangle, the cosine of an angle is equal to the ratio adjacent : hypotenuse or adjacent ÷ hypotenuse.

$a : h \quad \frac{a}{h}$

Counter-example An example which breaks the rule; the rule 'all prime numbers are odd' has a counter-example in the number 2 which is prime but even.

Cross-section The flat shape you get when you slice open a solid.

Cube (number) The result of multiplying a number by its square.

Cube (solid shape) A solid with six square faces.

Cube root The number which, when cubed, gives the number you started with, e.g. the cube root of 8 is 2 ($\sqrt[3]{8} = 2$).

Cubed A number which has been multiplied by its square, e.g. 2 cubed is 8 and 3 cubed is 27 ($3^3 = 27$).

Cubic centimetre The volume of a cube which is one centimetre along each straight edge; a unit of volume, equivalent to one millilitre of liquid.

Cubic metre The volume of a cube which is one metre along each straight edge; one cubic metre of water has a mass of one tonne.

Cubic millimetre The volume of a cube which is one millimetre along each straight edge; about the size of a grain of sugar.

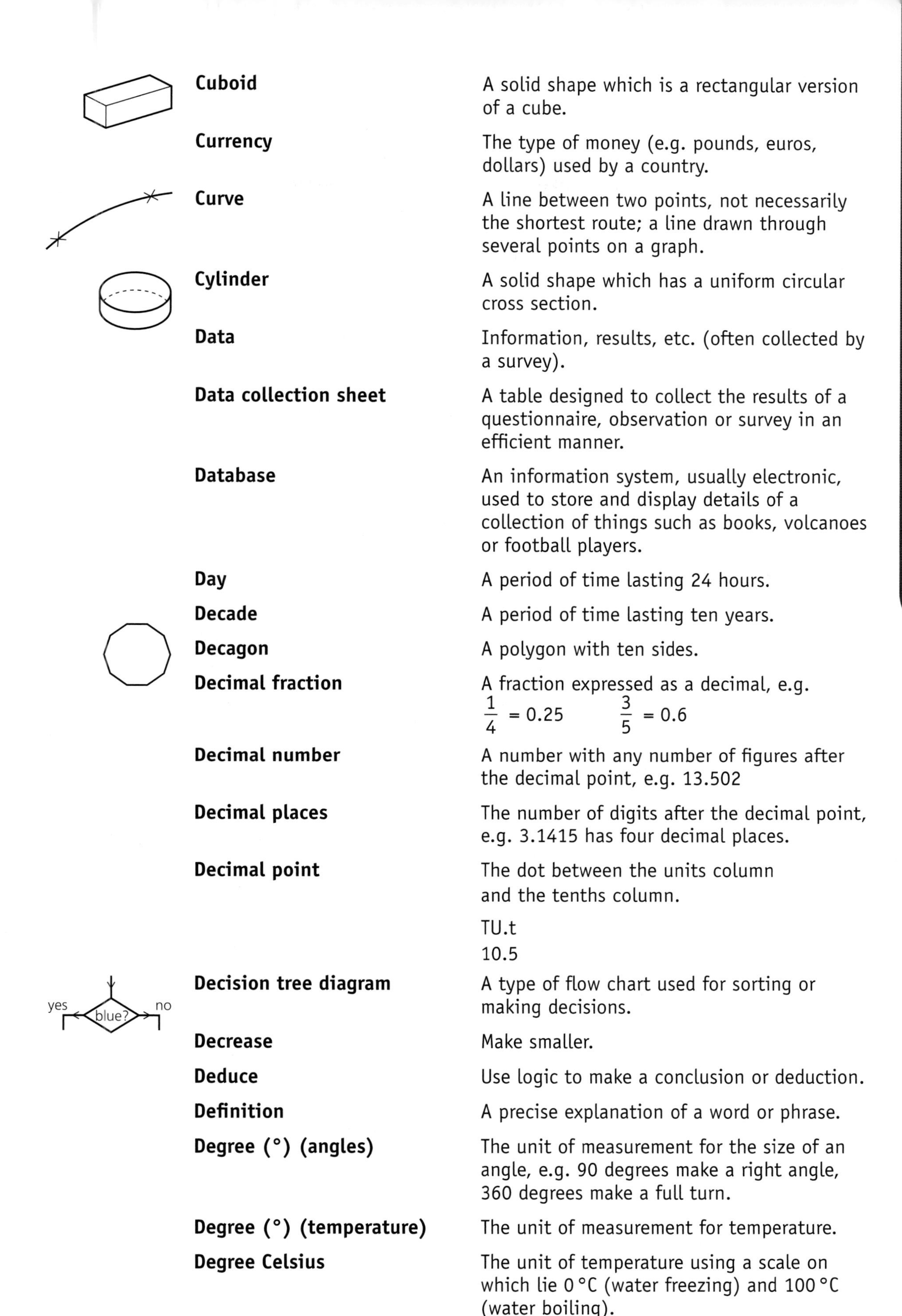

Cuboid A solid shape which is a rectangular version of a cube.

Currency The type of money (e.g. pounds, euros, dollars) used by a country.

Curve A line between two points, not necessarily the shortest route; a line drawn through several points on a graph.

Cylinder A solid shape which has a uniform circular cross section.

Data Information, results, etc. (often collected by a survey).

Data collection sheet A table designed to collect the results of a questionnaire, observation or survey in an efficient manner.

Database An information system, usually electronic, used to store and display details of a collection of things such as books, volcanoes or football players.

Day A period of time lasting 24 hours.

Decade A period of time lasting ten years.

Decagon A polygon with ten sides.

Decimal fraction A fraction expressed as a decimal, e.g.
$\frac{1}{4} = 0.25 \qquad \frac{3}{5} = 0.6$

Decimal number A number with any number of figures after the decimal point, e.g. 13.502

Decimal places The number of digits after the decimal point, e.g. 3.1415 has four decimal places.

Decimal point The dot between the units column and the tenths column.

TU.t
10.5

Decision tree diagram A type of flow chart used for sorting or making decisions.

Decrease Make smaller.

Deduce Use logic to make a conclusion or deduction.

Definition A precise explanation of a word or phrase.

Degree (°) (angles) The unit of measurement for the size of an angle, e.g. 90 degrees make a right angle, 360 degrees make a full turn.

Degree (°) (temperature) The unit of measurement for temperature.

Degree Celsius The unit of temperature using a scale on which lie 0 °C (water freezing) and 100 °C (water boiling).

Degree Fahrenheit The unit of temperature using a scale on which lie 32 °F (water freezing) and 212 °F (water boiling).

Degree of accuracy — A way of describing the reliability of a measurement, e.g. ‘to the nearest centimetre’.

Delta — Another name for an arrowhead or concave kite.

Denominator — The bottom number in a proper, improper or algebraic fraction. $\frac{3}{\mathbf{4}}$ $\frac{8}{\mathbf{5}}$ $\frac{ab}{\mathbf{3c}}$

Density — A way of comparing the masses of the same volume of two substances; density = mass ÷ volume, e.g. water has a density of 1 g/cm^3, iron has a density of approx. 7 g/cm^3, lead has a density of approx. 12 g/cm^3.

Deposit — A payment made at the beginning of a deal, which can either be returned later or count as a first payment.

Depth — How deep something is; usually a vertical measurement of distance starting at zero and increasing downwards.

Descending — Going down; numbers in decreasing order.

Diagonal — A line across a plane shape joining two vertices (corners).

Diagram — A picture used to illustrate a situation, especially to make a complex problem clearer.

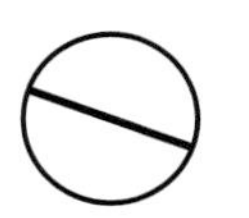

Diameter — A chord through the centre of a circle, cutting the circle in half.

Dice, die — One die, two dice: the (usually) cube-shaped solid used in games and probability experiments.

Difference — The answer to a subtraction.

Difference of two squares — A useful result in algebra which makes factorisation easier, $x^2 - y^2 \rightarrow (x + y)(x - y)$.

Difference pattern — The way in which the differences between terms of a sequence change as the sequence continues; this process, known as the ‘difference method’ will show up almost every hidden pattern in a sequence.

Digit (1) — One of the symbols from 0 to 9 used in writing numbers.

Digit (2) — Another word for a finger, especially when used to help in counting.

Dimension — A measurement of length, width or height.

Directed numbers — Numbers with a positive ($^+$) or negative ($^-$) sign; remember that +3 means ‘add 3’ and $^+3$ is a number or position on a number line; –3 means ‘subtract 3’ and $^-3$ is a number or position on a number line. (See also Sign and Sign change key.)

Direction	The way to be regarded as 'forwards', especially using the points of a compass or a three-figure bearing.
Discount	A reduction in the usual price.
Discrete	A measurement which can only take certain values, going up in 'jumps'.
Display (calculator)	The area on a calculator used to show the numbers entered and the calculation result.
Distance	The length of a line between two points.
Distance–time graph	A graph used to illustrate a journey.
Distribution	The way in which a set of data falls into groups in a frequency table.
Divide	Share; the first number is the one being shared.
Dividend (1)	A big number which is being divided up by a smaller number (divisor).
Dividend (2)	A reward paid out to shareholders (investors) by a company when it wants to share out the profits.
Divisibility	The property of being divisible by a number other than one.
Divisible	Can be divided by, e.g. 12 is divisible by 4
Division (1)	The operation of sharing one number equally into smaller parts.
Division (2)	The interval on a measuring scale.
Divisor	When 24 is divided by 3 to get 8, 3 is called the divisor (24 is the dividend and 8 is the quotient; there is no remainder).
Dodecagon	A polygon with 12 sides; a regular dodecagon has 12 equal sides and 12 equal angles.
Dodecahedron	A solid with 12 faces.
Double (noun)	The result of multiplying by two, as in '6 is the double of 3'.
Double (verb)	Multiply by two.
Doubt	Includes the possibility of failure as well as success.
Draw (1)	To draw in mathematics requires pencil, compasses and ruler.
Draw (2)	An even score at the end of a game.
Draw (3)	Select a card (at random) from a pack.
East	Towards the direction of the rising sun, usually shown as towards the right on a map.

distance

time

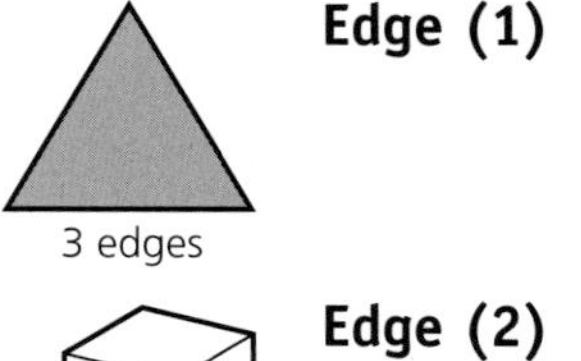

Edge (1)
The line joining two adjacent vertices in a polygon (also called a side); 'all the way around the edge' refers to the perimeter of the shape.

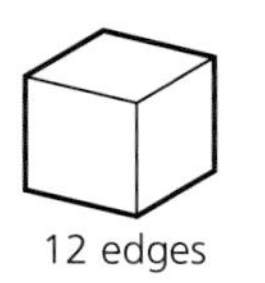
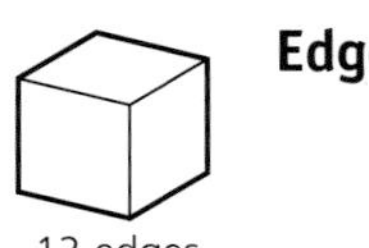

Edge (2)
The line shared by two adjacent faces in a 3D shape, running from one vertex to another.

Eighths
The result when a whole is divided into eight equal pieces.

Elements of a set
Members of a group which share the same property, e.g. in the set {vowels} are the elements {a, e, i, o u}.

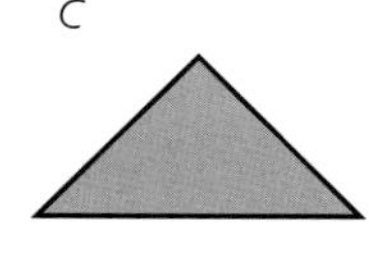

Elevation
The view of a three-dimensional solid from either the front, back or sides.

Enlarge
Make bigger.

Enlargement
A transformation which changes the size of a shape to make a similar image.

Enter (calculator)
Type in a calculation or command by a sequence of keystrokes; the final button pressed to obtain a result.

Equal angles
One of many situations in geometry in which two or more angles in a diagram are the same.

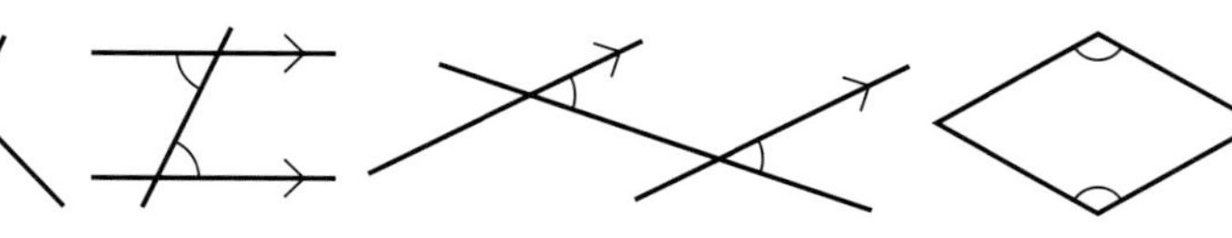

Equal sides
Sides which are the same length, as in isosceles or equilateral triangles, many quadrilaterals and all regular shapes.

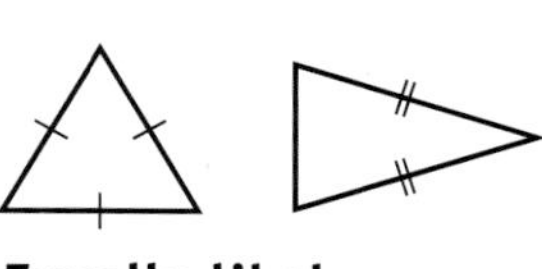
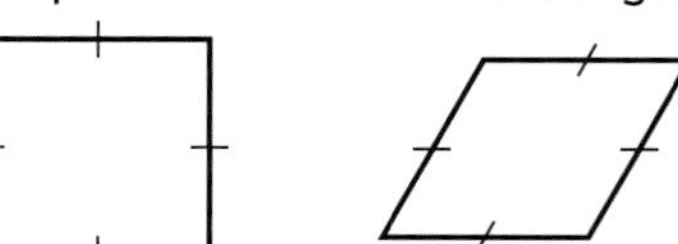
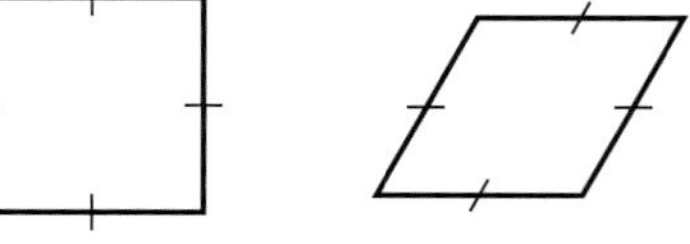

Equally likely
Having the same chance or probability.

Equals (=)
The expressions on either side have the same value.

Equation
An expression with '=' in it, usually to be solved.

Equidistant
The same distance away.

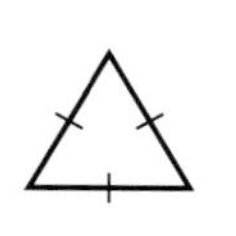
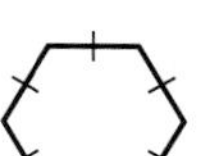
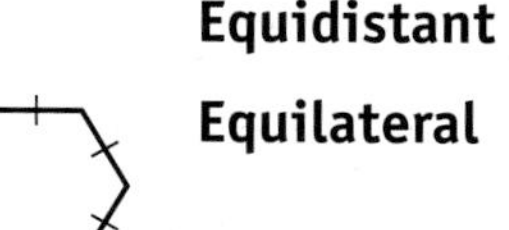

Equilateral
All sides of equal length, e.g. equilateral triangle.

Equivalence
Even better than being equal, e.g. $8x = 40$ is only true when $x = 5$; equivalence, however, is always true; e.g. $8x \equiv 5x + 3x$

Equivalent
Two things which are the same in every way, especially fractions and ratios, e.g.; $\frac{2}{4} \equiv \frac{4}{8}$ $4:12 \equiv 1:3$

Estimate Make a reasonable guess at the answer without detailed calculation.

Evaluate Work out exactly, using a reliable method of calculation.

Even A number which ends in 0, 2, 4, 6 or 8

Even chance The probability of 50%, as in 'heads or tails'.

Event One of any number of possible outcomes or future happenings.

Evidence Supporting material for a generalisation, but evidence alone does not make a proof.

Exact No further accuracy is possible.

Exactly The true answer, with no inaccuracies or approximations.

Exchange rate The figures used to convert one currency into another.

Exhaustive Every possibility has been listed.

Experiment Any trial which has at least one outcome.

Experimental probability For example, the experimental probability of getting a head on a coin is (number of heads counted) ÷ (number of times coin was tossed); this gets closer to the theoretical probability the more trials are done.

Explain Give a reason for something, especially using the word 'because' in the answer.

Explore Investigate an unknown area, problem or activity.

Expression A collection of terms, usually algebraic, e.g. $3x + 4 + 2x - 3$

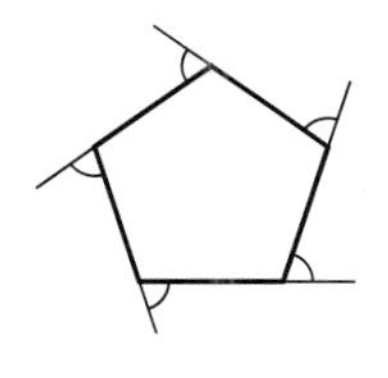

Exterior angle The angle through which a wire is bent at each vertex, when making a polygon from a straight piece of wire.

Face A flat surface of a solid shape, e.g. a cube has six faces.

Factor A small number which divides exactly into a bigger number, e.g. 4 is a factor of 12

Factor pairs Two numbers which multiply to give the required product, e.g. (3, 20) is a factor pair of 60

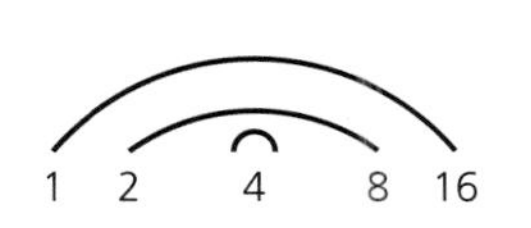

Factor rainbow Factors listed in order and joined in pairs can create a rainbow picture; if there is only one number in the middle then the original number was a square number, e.g. 16

Factorise	Rewrite a number or algebraic expression as a product of two or more factors; numbers are written as a product of primes, while expressions are written with one or more brackets. $36 = 3^2 \times 2^2$ $3y^3 = 3 \times y \times y \times y$ $3a + 6 = 3(a + 2)$
Fahrenheit	The temperature scale on which water freezes at 32 degrees and boils at 212 degrees.
Fair	Not biased; a fair die has an equal chance of landing on any one if its six faces.
False	Not true; a lie; having a logical value of zero (on a spreadsheet).
Fifths	The result when a whole is divided into five equal pieces.
Fifty-fifty chance	Another expression for evens or 50% probability.
Finite	Having an end eventually.
Flow chart	A chain of decision boxes or instruction boxes, useful for creating a logical process or organisation of ideas.
Fluid ounce (fl. oz)	An imperial measure of liquid capacity.
Fold (1)	Bend something over onto itself to double the thickness or make a crease.
Fold (2)	Suffix meaning 'many times' as in three-fold.
Foot	A unit of length consisting of 12 inches (about 30.48 cm) (pl. feet).
Formula	A 'recipe' for combining numbers, using words or letters (pl. **formulae**), e.g. $P = 2(l + w)$
Formulae	More than one formula (also known as formulas).
Fraction	A quantity written as a numerator over (divided by) a denominator, e.g. $\frac{2}{3}$ (proper) $\frac{3}{2}$ (improper).
Fraction wall	A way of comparing fractions by splitting the same width into different equal quantities.
Fractional shape	A shape formed from a simpler shape by cutting parts away.
Frequency	The number of times something has happened.
Frequency diagram	A type of bar chart in which the vertical axis is frequency, indicated by the height of the bar.

Frequency table/chart	A table of collected data organised into categories and giving the total (frequency) in each category.
Function	A rule for changing one number into another, e.g. 'multiply by 4' or 'add 1'.
Function machine	A chain of one or more functions which features an input and output of numbers.
Gallon	A unit of capacity consisting of 8 pints (about 4.5 litres).
General term	The term in a sequence at any desired position, also known as the nth term.
Generate	Produce, or continue, a sequence using the given rule.
Geometry	The branch of mathematics dealing with shapes.
Good chance	Having a probability above 50%.
Gradient	A measurement of the steepness of a line, given by (height increase) ÷ (width increase).
Gram	One thousandth part of a kilogram.
Graph	A pictorial representation of data.
Greater than (>)	Having a bigger value (note carefully which way the sign goes); the biggest number goes next to the big end of the sign, e.g. 5 > 3
Greater than or equal to (≥)	Having a bigger value or just the same value, e.g. if $x \geq 2$ for the score on an ordinary die, then x could be 2, 3, 4, 5 or 6
Greatest value	The maximum value obtained by a set of data or a line on a graph.
Grid	A network of crossed lines, usually at right angles to each other.
Grouped data	Data which has been sorted into groups (classes) of equal class interval.
Half	The result when a whole is divided into two equal pieces (pl. halves).
Halve	Divide by two.
HCF	See Highest Common Factor.
Hectare	A unit of area equal to 10 000 square metres (think of a square 100 m on each side); approx. 2.5 acres.
Height	The vertical distance between the top and the ground or base.
Hemisphere	Half of a sphere.
Hendecagon	A polygon with 11 sides.

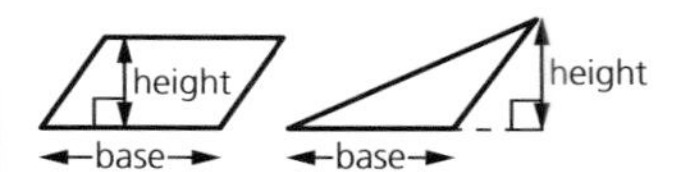

Heptagon A polygon with seven sides.

Hexagon A polygon with six sides.

High Having a large value; a measurement of the height of something.

Highest Common Factor (HCF) The biggest number which divides exactly into two or more larger numbers, e.g. 8 is the HCF of 16 and 24

Histogram A type of bar chart where the area of each bar gives the frequency.

Horizontal Parallel to the horizon; straight across.

Hour A unit of time equal to 60 minutes or one 24th of a day.

Hundred, hundreds The place value position to the left of the tens column. **H** T U.t h

Hundredth, hundredths The place value position to the right of the tenths column. H T U.t **h**

Hundredweight One twentieth of an imperial ton (112 lb).

Hypotenuse The longest side in a right-angled triangle.

h

Icosahedron A regular solid with 20 equilateral triangle faces.

Identical Exactly the same in all respects; congruent (shapes).

Identically equal to Another way of saying 'equivalent to', i.e. always true.

Image The result of a transformation on an object.

Imperial The measurement system in use before the metric system.

Impossible Not possible under any circumstance; probability 0; guaranteed to fail.

Improper fraction A fraction in which the numerator (top) is bigger than the denominator (bottom) – can be changed into a mixed number e.g.

$\frac{3}{2} \rightarrow 1\frac{1}{2}$ $\frac{15}{4} \rightarrow 3\frac{1}{4}$

In terms of A formula in terms of t will use only the variable t combined with numbers.

Inch Imperial unit of length (based originally on the length of part of a Roman thumb!).

1 inch

Increase Make bigger.

Index The little number which indicates repeated multiplication, squares, cubes, etc. e.g. 2^3

Index notation Writing numbers in a form which shows the correct power of 10 each time, e.g. 3000 is 3×10^3

Indices One index, two indices; see also Power.

Inequality A statement featuring 'less than' or 'greater than'.

Infinite Having no end, ever; you can never reach infinity.

input × 2 output

Input A number which is 'fed into' a function machine; see also **Function**.

Instruments Mathematical equipment such as compasses and ruler.

Integer A whole number such as 35 or 108 or $^{-}2$

Integer solution set All the whole numbers which satisfy a pair of inequalities, e.g. $x \le 2$ and $x \ge {}^{-}2$, which gives the integer solution set $\{^{-}2, {}^{-}1, 0, 1, 2\}$

Intercept 'Cut across'; a useful word to describe where a straight line crosses the y-axis.

Interest Money paid into a savings account by a bank or building society; usually a percentage of what was there before (compound interest).

Interior angle An angle between two adjacent sides inside a polygon; in a regular polygon all the interior angles are equal.

Interpret Give meaning to something, especially a set of results or a diagram.

Intersect (1) To cross, especially in the context of straight lines.

A B

Intersect (2) To overlap, especially in the context of sets.

Intersection The point at which two intersecting lines meet; this point of intersection is on both lines.

Interval Any part of a scale, especially on a number line.

5 × 3 15
5 ÷ 3 15

Inverse The opposite, e.g. × and ÷, + and –; square and square root.

+ 3
– 3

Inverse function The function which does the opposite of the one given, e.g. '– 3' (minus 3) is the inverse function of '+ 3' (plus 3).

Inverse mapping The reverse mapping, e.g. $x \rightarrow x - 3$ is the inverse mapping for the mapping $x \rightarrow x + 3$

Investigate Find out by asking one's own questions and looking for answers.

Irregular Not regular (shapes), thus having two or more different sides and angles.

Isometric A grid of equilateral triangles, used for drawing solid shapes in two dimensions.

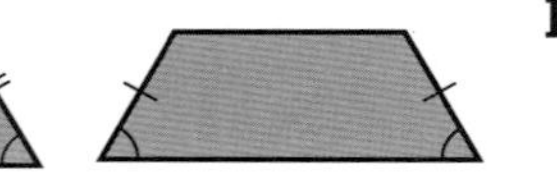

Isosceles Having two sides the same length to create a line of symmetry (especially isosceles triangle and isosceles trapezium).

Justify Give a reason for something, e.g. algebra as a powerful way to justify a statement such as 'odd + odd = even'.

Key (1) Any one of the buttons on a calculator which can be pressed in the course of a calculation.

Key (2) An indicator to provide essential information on a graph or scale drawing, e.g. scale 1 cm : 100 m; or a colour code for a pie chart.

Kilogram One thousand grams; the standard unit for measuring mass.

Kilometre One thousand metres; used for measuring long distances.

Kilometres per hour (kph or km/h) Measure of speed obtained by dividing distance travelled (in km) by time taken (in hours).

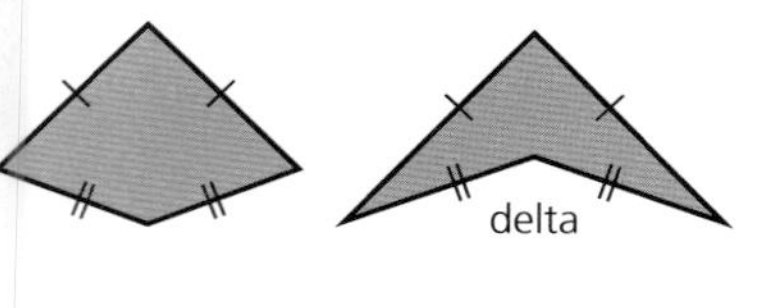

Kite A symmetrical quadrilateral with two adjacent pairs of equal sides.

Label The description of an axis on a graph or chart, e.g. 'y-axis' or 'frequency'.

LCM See Lowest Common Multiple.

Least significant digit The digit in a number which has least say in the size of the number, e.g. the 7 in 246.7

Least value The minimum value attained by a set of data or a line on a graph.

Length The distance between the two ends.

Less than (<) Having a smaller value; note carefully which way the sign goes; the biggest number goes next to the big end of the sign, e.g. $3 < 5$
If $x < 3$ then x could be … $^{-}2$, $^{-}1$, 0, 1, 2

Less than or equal to (≤) Having a smaller value or just the same value. If $x \leq 3$ then x could be …
$^{-}2$, $^{-}1$, 0, 1, 2, or 3

Like terms Algebraic terms with the same letter part; these can be added or subtracted, e.g. $3x^2$, x^2 and $^{-}2x^2$ are all like terms.

Likelihood Chance; e.g. in the phrase: 'the likelihood of winning the Lottery is very small'.

Likely Describing something which has a good chance of happening.

Limit The boundary beyond which the value cannot go; when you divide any number by the number just after it, you can get as close as you like to 1 but you will never quite get there; 1 is the limit.

Line Originally the shortest distance between two points; in general, a line has infinite length but no breadth.

Line graph A graph drawn by joining up plotted points (see also Time graph).

Line of best fit A straight line judged by eye to go through the majority of points on a scatter graph, although it may not go through any of them!

Line of symmetry One of the mirror lines on a diagram with reflection symmetry.

Line segment A piece of an infinite line, with two end points.

Line symmetry Another name for reflection or reflective symmetry.

Linear Like a straight line.

Linear equation An equation containing only letters and numbers i.e. no squares or cubes of letters, e.g. $3x + 4 = 2x + 7$

Linear expression A type of expression with no squares or higher powers, e.g. $3x + 5$

Linear function A function which involves only the operations +, −, × and ÷ on the input number, which can be represented by a straight line, e.g. $y = 3x + 5$

Linear relationship A mapping which uses only linear functions.

Linear sequence A sequence of numbers which, if plotted as y co-ordinates for consecutive x values, would lie on a straight line; going up (or down) by the same amount each time, e.g. 15, 12, 9, 6 ...

Litre One thousand millilitres; used for measuring quantities of liquid.

Long division A pencil and paper method of dividing one number by another.

Long multiplication A pencil and paper method of multiplying two numbers.

Loss A 'negative profit', where the money received in sales is less than the value of the goods sold.

Lowest common multiple (LCM) The smallest number into which two or more smaller numbers will divide exactly, e.g. the LCM of 6 and 9 is 18. The Lowest Common Denominator is an example of an LCM.

Lowest terms Describes a fraction which cannot be cancelled down (simplified) any more, e.g. $\frac{1}{3}$ $\frac{3}{7}$ $\frac{10}{21}$

Map (1) A scaled picture or scale diagram of an area.

Map (2)/mapping — A procedure for changing one number (input) to another (output).

Map (3) — The transformation of a shape mapped onto an image by means of reflection, rotation or translation.

Mass — A measure of the amount of matter; the 'weight' of an object, measured in grams and kilograms; see also **Weight**.

Maximum value — The greatest size a number can take in the given context.

Mean — The average found by adding the data and dividing the total by the number of items, e.g. the mean of 1, 2, 3 and 18 is 6

Measure — The value or size of something using standard units.

Median — The average found by writing data in order and finding the middle value, or the mean of the two middle values, e.g. the median of 1, 2, 3 and 14 is 2.5

Memory — The facility for storing numbers and half-way answers to assist in a calculation, when using a calculator.

Mensuration — The process of measuring and calculating measurements.

Method — A procedure, process or 'recipe' for carrying out a calculation or operation, e.g. the method for constructing triangles needs a ruler, protractor and compasses.

Metre — The standard unit of length, equal to one hundred centimetres or one thousand millimetres.

Metres per second (m/s) — A unit of speed in which one metre is covered every second.

Metric — The measurement system based upon powers of 10 with the metre as the basic unit of length.

Metric/imperial approximations — Rough equivalents used to convert quickly between units in the metric and imperial systems.

Midpoint — The half-way point along a line, e.g. the point of intersection of the diagonals of a parallelogram.

Mile — A unit of length corresponding to 'a thousand paces' (5280 feet) or about 1609 metres (5 miles is approximately 8 km).

Miles per hour (mph or miles/h) — A unit of speed in which one mile is covered every hour.

Millennium — A period of time lasting 1000 years.

Millilitre	One thousandth part of a litre.
Millimetre	One thousandth part of a metre.
Minimum value	The smallest size a number can take in the given context.
Minuend	The number from which another (the subtrahend) is subtracted.
Minus (negative)	A signed or directed number which is below zero; better known as 'negative numbers'.
Minus (subtract)	Take away; the second number is taken away from the first.
Minute (1) (time)	A period of time lasting 60 seconds or one 60th of an hour.
Minute (2)	Very small.
Mirror line	A line through the middle of a shape where one could place a mirror and still see all of it.
Mixed number	A number with a whole number part and a proper fraction part; can be converted into an improper fraction for multiplying and dividing calculations, e.g. $1\frac{1}{2} \rightarrow \frac{3}{2}$ $3\frac{1}{4} \rightarrow \frac{13}{4}$
Mnemonic	An aid to remembering, e.g. 'Naughty Elephants Squirt Water' is a mnemonic for North East South West (remembering to work clockwise!).
Modal class/group	The class/group in a table of grouped data which has the highest frequency.
Mode	Occurring most often; most frequent or most common; the average represented by the most commonly occurring data item.
Month	A period of time lasting approximately four weeks (lunar month) or 28–31 days (calendar month).
Most significant digit	The digit in a number which has greatest say in the size of the number, e.g. the 2 in 246.7
Multiple	A product, e.g. multiples of 3 are all the numbers (including 3) with 3 as a factor, i.e. numbers in the 3-times table.
Multiplicand	The number which is being multiplied by something else (the multiplicator).
Multiplication	The process of finding the product; i.e. multiplying.
Multiplicator (multiplier)	The number which is being used to multiply up another number (the multiplicand).
Multiply	Find the product of two or more numbers; 'times', as in the times tables.

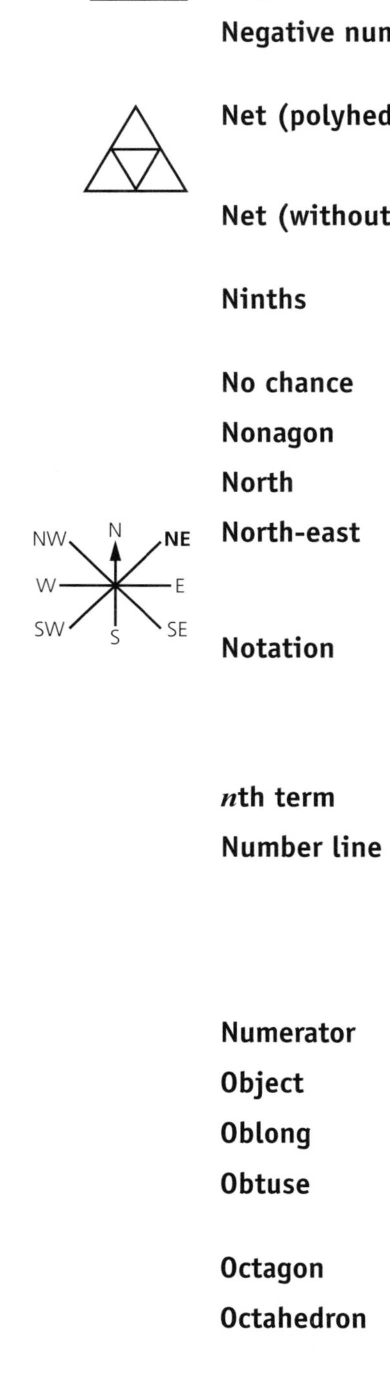

Multiply out	Multiply each term inside the brackets by the factor outside and then simplify the resultant expression, e.g. $3(x + 4) = 3x + 12$ $x(x - 3) = x^2 - 3x$ $2x(2x + 1) = 4x^2 + 2x$
Nearest	Closest, especially when rounding numbers, e.g. 3467 is roughly 3470 to the nearest 10, or roughly 3500 to the nearest 100
Nearly	Almost, as in 'the value of pi is nearly 3 and one-seventh'.
Negative	The opposite of positive; below zero, e.g. negative six is written as $^-6$
Negative correlation	As one variable increases, the other decreases.
Negative numbers	Numbers which are below zero, occasionally referred to as 'minus numbers', e.g. $^-3$
Net (polyhedron)	A 'cardboard cut-out' design which can be folded up to make a solid shape.
Net (without tax)	The value or price of something before VAT has been added on.
Ninths	The result when a whole is divided into nine equal pieces.
No chance	Zero probability; impossible.
Nonagon	A polygon with nine sides.
North	Towards the North Pole.
North-east	With north-west, south-west and south-east: the points of the compass in between the four main directions.
Notation	Using algebra to write down a rule or expression for the general term (nth term) of a sequence in a convenient form, e.g. $T(n)$ **or** Tn **or** T_n.
***n*th term**	The general term of a sequence of numbers.
Number line	Numbers in order on a marked line, usually either side of zero, (especially as an aid to mental or pencil arithmetic) for plotting points or for representing a solution set for a pair of inequalities.
Numerator	The top number in a fraction, e.g. the 3 in $\frac{3}{5}$
Object	The original shape before being transformed.
Oblong	A rectangle which is not square.
Obtuse	An angle greater than 90 degrees but less than 180 degrees.
Octagon	A polygon with eight sides.
Octahedron	A regular solid shape with eight equilateral triangle faces.
Odd	A number which ends in 1, 3, 5, 7 or 9

Operation — The four basic calculations done with numbers (+, −, ×, ÷).

Opposite (1) — Facing; on the other side; in a triangle, the side furthest from the angle.

Opposite (2) — An inverse, e.g. multiplication is the opposite of division.

Opposite angles (1) — When two lines cross, they make an X shape: the angles above and below the point of intersection are opposite and equal angles; left and right are also opposite and equal. Opposite angles are often called 'vertically opposite angles'.

Opposite angles (2) — Two interior angles on opposite sides in a regular polygon; they can be joined by a diagonal which cuts the polygon in half.

Opposite sides — Sides in a regular polygon which are separated by the width of the shape – the regular polygon centre lies between them; a regular polygon with an odd number of sides has no pairs of opposite sides.

Order (1) — To list, e.g. a set of numbers according to size.

Order (2) — The rules that say which of the various operations (e.g. +, −, ×, ÷) are to be given priority; also known as **BODMAS** or **BIDMAS**.

Order (of rotation symmetry) — The number of ways in which a plane shape appears the right way up (maps onto itself) while being turned through 360 degrees.

order 2

order 3

Origin — The point with co-ordinates (0, 0) on an (x, y) grid.

Original value — The size of something at the beginning, commonly before a percentage change.

Ounce — A unit of mass (abbreviated oz) which represents one-sixteenth of a pound weight (about 28 g).

Outcome — A possible result of a particular event.

Outcome table — A systematic way of displaying all possible outcomes to an experiment.

Output — A number which 'comes out' of a function machine.

P(n) — For the probability of an event n, e.g. P(heads) is shorthand for 'the probability of getting heads'.

Parallel — Running in the same direction.

Parallelogram	A quadrilateral with two pairs of parallel sides.
Partition	Separating a number into its parts, e.g. 126 = 100 + 20 + 6 (the hundreds, tens and units parts).
Pattern (1)	An arrangement of objects, e.g. matches, counters or tiles, according to some rule which enables a sequence to be made.
Pattern (2)	A pattern of numbers enabling us to predict the next number in the sequence.
Pattern (3)	A geometrical design which is built by applying rules of symmetry.
Pentagon	A polygon with five sides.
Per	Divided by; for each one; out of; as in per cent or metres per second.
Percentage (%)	A number which is a fraction of 100, e.g. $30\% = \frac{30}{100}$
Perimeter	The distance all the way around the edge of a shape.
Perpendicular	At right angles.
Perpendicular bisector	A line which cuts another line in half, meeting it at right angles; the construction using a ruler and compasses which creates this line.
pi (π)	The ratio of a circle's circumference to its diameter, approximately 3.14
Pictogram	A diagram similar to a bar chart in which appropriate pictures are put in lines to make the (usually) horizontal bars.
Pie chart	A circular diagram in which the size of each sector shows the proportion of the whole.
Pint	A unit of capacity (usually liquid) consisting of 20 fluid ounces (about 0.6 litres).
Place value	Units, Tens, Hundreds etc; the size of a number depends on its position. H T U.t h
Plan	A bird's eye view of a solid, especially used to show floor layouts in buildings.
Plan view	The view from above.
Plane	A flat surface; two-dimensional, like a sheet of paper.
Plane shapes	Two-dimensional shapes which are drawn on paper, as opposed to solid shapes, which can be picked up.
Plane symmetry	The three-dimensional equivalent in solids of line symmetry in plane shapes.

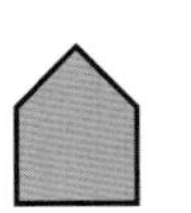

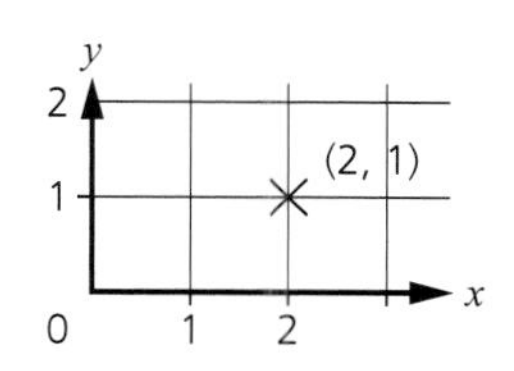

Plot	Mark a small cross to indicate a position of a point.
Plus	The arithmetical sign (+) to indicate addition.
p.m.	*post meridiem* (Latin) which means 'after noon'; see also **a.m.**
Point	A position often indicated by a cross; having no dimension.
Polygon	A shape with straight sides.
Poor chance	Having a probability below 50%.
Position	Another name for place or location; in mathematics we describe positions precisely using co-ordinates.
Positive	Above zero on the number line, e.g. positive three is written as $^{+}3$ (positive 3) or simply 3
Positive correlation	As one variable increases, so too does the other.
Possibility	One of the things which could happen.
Possibility space	A table listing systematically all possible outcomes of an experiment.
Possible	Having a probability which is greater than 0
Pound (money)	£1; equal to 100 pence.
Pound (weight)	1 lb equal to 16 oz (about 0.5 kg).
Power	The top right (index) number used in squares, cubes etc. to indicate repeated multiplying, e.g. 2^3 means 2 to the power of 3 (i.e. $2 \times 2 \times 2$).
Predict	Guess the outcome of a future event, usually choosing the outcome with the highest probability.
Prime	A number which has only two factors, 1 and itself.
Prime factor	A factor which is prime; a number which divides exactly into a larger number but which does not itself have any smaller factors besides itself and 1, e.g. the prime factors of 14 are 2 and 7
Prime factor decomposition	Writing a composite number as a product of its prime factors, e.g. $180 = 2 \times 2 \times 3 \times 3 \times 5$
Prism	A solid which has the same shaped cross-section all the way through.
Probability	The branch of mathematics which looks at the chance of things happening.

Probability scale The number line from 0 (impossible) to 1 (certain).

0 — 1
impossible — unlikely — even chance — likely — certain

Probability sum It is always certain that something will happen, so the sum of all probabilities at any time is 1

Probability tree The branching diagram used to keep track of multiple-event probabilities.

Probable Another name for possible, but usually used to mean with a probability greater than 50%.

Problem Any kind of question to which an answer is needed.

Procedure A sequence of instructions.

Product The result of a multiplication.

Product of prime factors A way of writing any whole number as a product of its component primes, e.g. $60 = 2 \times 2 \times 3 \times 5$

Profit The overall gain made by the seller when something is sold; profit = selling price – cost price.

Proof A logically strong argument which explains why something is always true (note that an example does not make a proof!).

Proper fraction A fraction in which the numerator (top) is smaller than the denominator (bottom) – when written in decimal form it will be between 0 and 1, e.g. $\frac{3}{4} = 0.75$

Property A characteristic of something, e.g. one of the properties of a kite is that it has one line of symmetry.

Proportion Any of several ways (e.g. fraction or percentage) used to describe part of a whole, e.g. 'the proportion of girls in this room is about three-fifths'.

Proportional to (in proportion to) Where multiplying one of two numbers by another number requires the second to be multiplied by the same number, e.g. adapting a recipe for a different number of people.

Proportionality The property of being in proportion.

Protractor Geometrical instrument used to measure angles, commonly made of a semicircle of clear plastic.

Prove Provide an argument which leaves no room for doubt.

Pyramid A solid with a flat (esp. square) base and the other faces sloping up to a point at the apex.

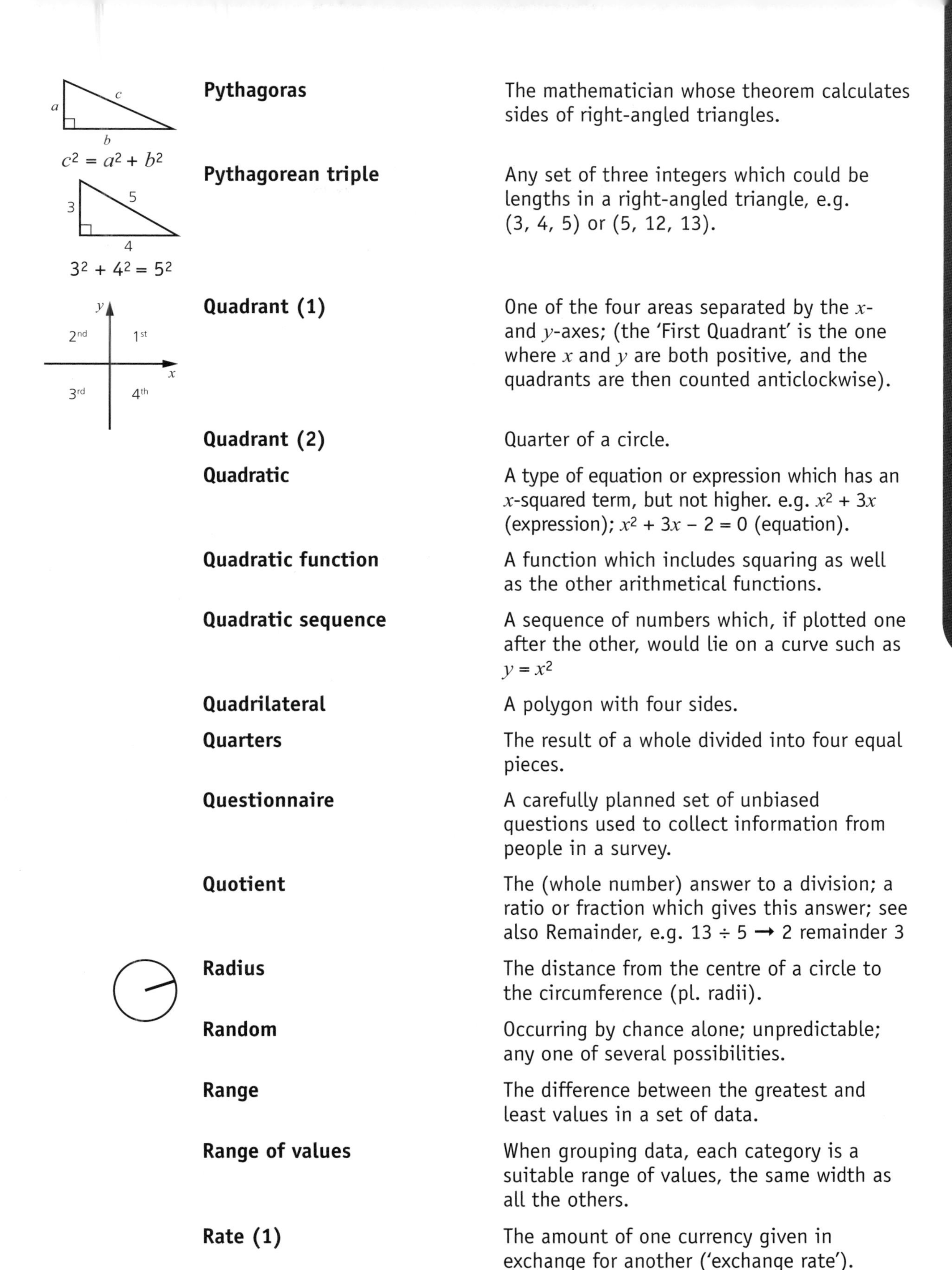

Pythagoras	The mathematician whose theorem calculates sides of right-angled triangles.
Pythagorean triple	Any set of three integers which could be lengths in a right-angled triangle, e.g. (3, 4, 5) or (5, 12, 13).
Quadrant (1)	One of the four areas separated by the x- and y-axes; (the 'First Quadrant' is the one where x and y are both positive, and the quadrants are then counted anticlockwise).
Quadrant (2)	Quarter of a circle.
Quadratic	A type of equation or expression which has an x-squared term, but not higher. e.g. $x^2 + 3x$ (expression); $x^2 + 3x - 2 = 0$ (equation).
Quadratic function	A function which includes squaring as well as the other arithmetical functions.
Quadratic sequence	A sequence of numbers which, if plotted one after the other, would lie on a curve such as $y = x^2$
Quadrilateral	A polygon with four sides.
Quarters	The result of a whole divided into four equal pieces.
Questionnaire	A carefully planned set of unbiased questions used to collect information from people in a survey.
Quotient	The (whole number) answer to a division; a ratio or fraction which gives this answer; see also Remainder, e.g. $13 \div 5 \rightarrow 2$ remainder 3
Radius	The distance from the centre of a circle to the circumference (pl. radii).
Random	Occurring by chance alone; unpredictable; any one of several possibilities.
Range	The difference between the greatest and least values in a set of data.
Range of values	When grouping data, each category is a suitable range of values, the same width as all the others.
Rate (1)	The amount of one currency given in exchange for another ('exchange rate').
Rate (2)	The measure of progress or increase with time, metres per second; litres per hour, etc.
Ratio	A way of comparing the relative size of two or more whole numbers, e.g. 3 : 2
Raw data	Data as gathered in a survey, before being 'processed' by grouping, averages, graphs, etc.

Reason An explanation given in answer to the question 'why?'.

Reasoning The process of clear thinking which uses logic and careful explanation.

Record (1) (noun) An entry in a database.

Record (2) (verb) To gather information in a survey and write it down carefully in a table.

Rectangle A quadrilateral with four right angles; see also **Oblong**.

Recurring decimal A fraction converted into decimal form which continues in a repeated pattern for ever, e.g. $\frac{2}{9} \rightarrow 0.222\ldots$

Reduce/reduction The opposite of increase/enlargement.

Reflect Draw the result of a reflection in a given mirror line.

Reflection (1) The image seen in a mirror or mirror line.

Reflection (2) The transformation which maps an object onto its reflected image.

Reflection symmetry The property of having matching halves either side of one or more mirror lines.

Reflex Describes an angle which is bigger than 180 degrees.

Region (1) A part of a surface, e.g. one marked out by lines on a grid or in a shape.

Region (2) Part of a Venn diagram which illustrates a subset of the universal data set.

Regular Having all the sides the same length and all the angles equal (regular polygons).

Relationship The connection between two or more numbers, including unknown numbers, often written as a formula.

Remainder Division gives a whole number part (quotient) and a remainder, either of which can be zero, e.g. $13 \div 5 \rightarrow 2$ remainder 3

Represent Give in another form, especially as a diagram or graph.

Result (1) The outcome of a survey or experiment, best displayed in a table.

Result (2) The final answer to a numerical problem or calculation.

Result (3) The effect of completing a transformation such as rotation or reflection.

Rhombus (diamond) A symmetrical quadrilateral which has four equal sides.

Right (1)	Correct.
Right (2)	Describes an angle size of 90 degrees.
Right-angled	With one or more angles of 90 degrees.
Risk	A gamble; an experiment where failure has a non-zero probability.
Roman numerals	I (1), V (5), X (10), L (50), C (100), D (500), M (1000).
Roman numbers	Numbers written in Roman numerals, examples MMXIII (2013), DCCI (701); sometimes, but not always, a shortened form is written, examples IV (4, '1 before 5'), XC (90, '10 before 100').
Rotate	Draw the result of a rotation through a given angle and about a given point.
Rotation (1)	Turning around a fixed point.
Rotation (2)	The transformation which maps an object onto its rotated image.
Rotation symmetry	The property of having two or more 'right ways up'.
Rough	Approximate.
Roughly	Not given to a high degree of accuracy, e.g. an approximation.
Round (1)	The shape of a circle.
Round (2)	Simplify or approximate an answer so that fewer significant figures are required to write it down, e.g. 4745 is 5000 to the nearest thousand, 4700 to the nearest hundred and 4750 to the nearest ten.
Row	The horizontal in a table (left – right).
Rule (1)	The instructions for generating a sequence of numbers.
Rule (2)	A straight line, as drawn by pencil and ruler.
Ruler	Any instrument for drawing straight lines or measuring length.
Sale price	The new price of an item after a reduction or discount has been applied.
Sample	A selection, preferably random and unbiased, for the purpose of a survey or experiment.
Sample space	An organised list of all the possible samples that could be made, also called an 'outcome' space.
Satisfies	A value satisfies an equation if the equation can be shown to be true when the value is substituted.

Scale (1) Measuring scales such as rulers, bathroom scales, electricity meters, etc.

Scale (2) A unitary ratio, e.g. 1 : 20 000 which means that 1 cm on the drawing represents 20 000 cm (200 m) in real life.

Scale (3) Used when drawing a graph; choose the right scale on the axes so that all the numbers can be plotted on the paper clearly.

Scale drawing A drawing accurately drawn from scaled measurements from the original.

Scale factor Used to reduce or enlarge the original, e.g. on a diagram or map.

Scalene All sides of different length, e.g. scalene triangle.

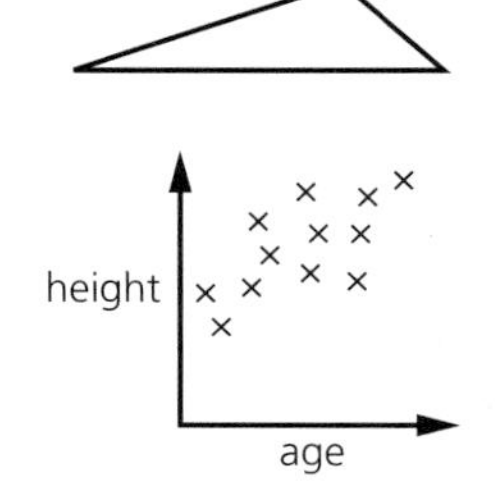

Scatter graph Points of paired data e.g. (age, height) plotted as (x, y) co-ordinates in order to test whether or not there is a relationship (correlation) between the two.

Sea level A zero reference height defined by the average of high and low tides, so that all other heights can be compared with it (above and below).

Second (1) The one after the first and before the third.

Second (2) A unit of time equal to one-sixtieth of a minute.

Section A piece, especially when referring to a part of a line or part of a shape; also used to mean 'cross-section'.

Sector Part of a circle cut out by two radii, e.g. 'a pie slice' in a pie chart.

Segment Part of a circle cut out by a chord.

Semicircle Half of a circle, cut out by a diameter.

Sequence (1) A list of numbers following a logical pattern or rule, e.g. 1, 3, 5, 7, 9, ...

Sequence (2) A series of patterns following a rule, e.g.

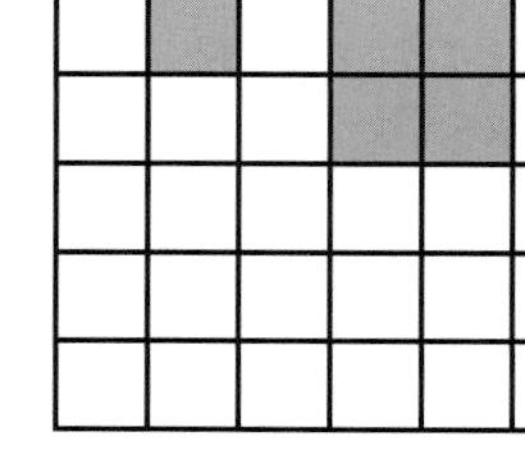

Service charge The extra charge on a bill to pay for services, e.g. in a restaurant, the charge for the waiter or waitress.

Set — A collection of things which all have something in common, e.g. {children who learn the piano}, {prime numbers}, {quadrilaterals}, etc.

Set notation — The collection of symbols used to describe things in sets.

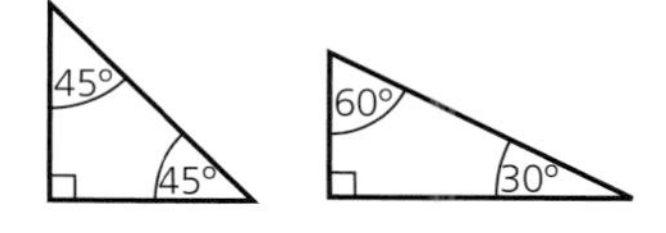

Set square — Triangular device used to draw/measure angles of (45, 45, 90) degrees or (30, 60, 90) degrees.

Sevenths — The result of a whole divided into seven equal pieces.

Shape — A geometrical figure; an enclosed space or form.

Share — Divide a given amount equally.

Side (1) — A straight line between the vertices in a geometric shape.

Side (2) — Reference to either left or right of the = sign in an equation.

Sign — The direction of a number, shown as $^{+}$ (positive, which is often omitted) or $^{-}$ (negative) in front of the number. e.g. $^{+}3$ ('positive 3', or just '3') $^{-}2$ (negative 2).

Sign change key — The key [+/−] or [(−)] on a calculator which changes the sign of a directed number.

Significant figures — Those figures which, in a form of rounding, are concerned more with the order of magnitude of the number than with the number of decimal places, e.g. 3473 is 3500 to 2 s.f., 3000 to 1 s.f. or 3470 to 3 s.f. 0.408 is 0.41 to 2 s.f. or 0.4 to 1 s.f.

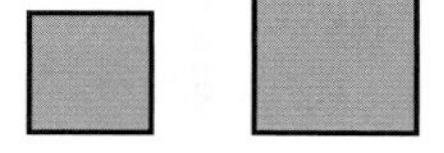
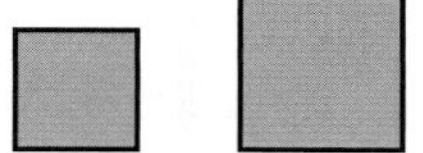

Similar — Having the same shape, but of a different size, e.g. an enlargement.

Similarity — The property of being similar, i.e. the same shape (not to be confused with its non-mathematical meaning of 'being nearly the same').

Simplest form — Most commonly used to refer to fractions after they have been cancelled by removing common factors above and below (lowest terms); also used for ratios.

e.g. $\frac{3}{6} \to \frac{1}{2}$ $\quad$ $9:12 \to 3:4$

Simplify — Reduce to a simpler form, especially with fractions, ratios and algebraic expressions, e.g. $4:2 \to 2:1$ $\quad$ $3x + 4x \to 7x$

Simultaneous equations — Two or more equations which need to be solved at the same time,

e.g. $\left.\begin{aligned} x + y &= 10 \\ x - y &= 4 \end{aligned}\right\}$

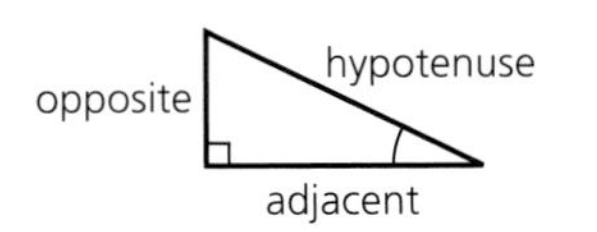

Sine (sin) In a right-angled triangle, the sine of an angle is equal to the ratio opposite : hypotenuse or opposite ÷ hypotenuse.

Sixths The result when a whole is divided into six equal pieces.

Sketch A diagram drawn more for convenience than accuracy.

Slope Gradient; sloping.

Solid A three-dimensional shape which can be picked up and handled, as opposed to plane shapes which can only be seen on paper.

Solution The answer to a problem, equation or inequality.

Solve Find the value of the variable, by a systematic method, which will make the equation correct.

South Towards the South Pole.

Speed The rate at which distance is travelled, commonly in metres per second or miles per hour.

Sphere A ball-shaped solid.

Spin Turn (usually fast) especially to determine an outcome, e.g. coin, spinner, etc.

Spinner Anything capable of spinning, especially one which can indicate a variety of possible outcomes.

Square (1) A regular quadrilateral.

Square (2) To multiply a number by itself; the result obtained when this is done, e.g. 3 × 3 = 9

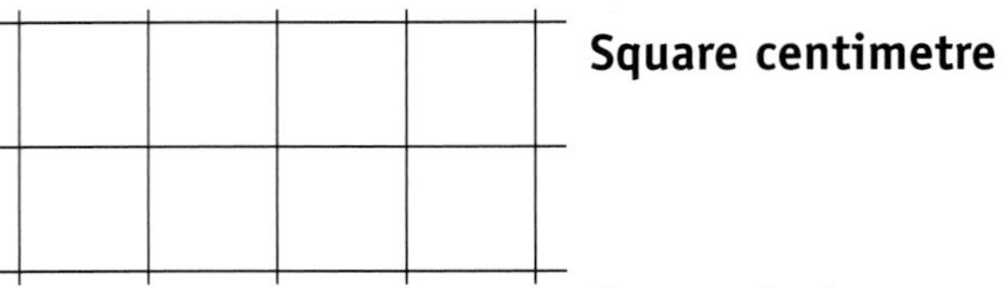

Square centimetre A square, one centimetre on each side; a unit of area for small shapes and regions, such as those drawn in exercise books.

Square inch A square, one inch on each side; an imperial unit for area occasionally seen in the measurement of pressures, e.g. pounds per square inch or p.s.i. as on some car tyres.

Square kilometre A square, one kilometre on each side; a unit of area for very large shapes and regions, such as countries.

Square metre A square, one metre on each side; a unit of area for large shapes and regions, such as playgrounds and tennis courts.

Square mile A square, one mile on each side.

Square millimetre A square, one millimetre on each side; a unit of area for very small shapes and regions, such as things seen under a microscope.

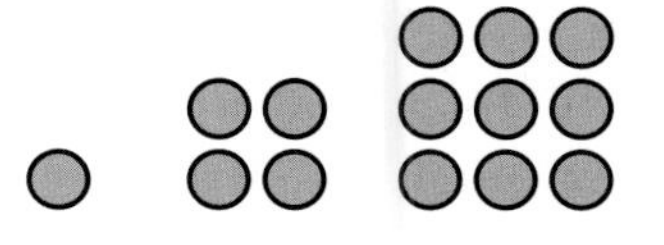

Square number The result of multiplying a whole number by itself; can be drawn as a square of dots, e.g. 1, 4, 9, 16, 25 etc.

Square root The number which when squared gives the one at the start, e.g. the square root of 64 is 8 ($\sqrt{64} = 8$)

Square yard A square, one yard on each side.

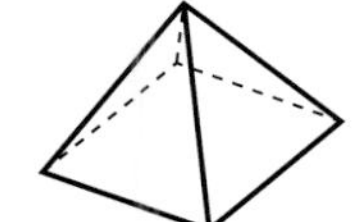

Square-based pyramid A solid shape with a square base, with the other four faces rising to a single point – the shape of the famous pyramids in Egypt.

Squared Multiplied by itself, as in three squared is nine (written as $3^2 = 9$).

Standard (index) form See Index notation.

Statistics The branch of mathematics dealing with understanding data.

Steepness A measure of the slope or gradient of a line; (very steep lines are almost vertical).

Stone A unit of weight equivalent, in Earth's gravity, to a mass of 14 pounds (6.35 kg).

Straight angle Two lines which meet at an angle of 180°, thus making a single straight line.

Straight edge Any kind of ruler or similar which is used only to assist with drawing diagrams and constructions.

Straight-line graph A graph of a linear function, such as $y = 5x + 2$

Subset A set of related objects which forms part of a larger set, for example: {girls in Year 7} is a subset of the set {girls in the school}.

Substitute Replace the letters in an algebraic expression with numerical values.

Substitution The process of changing letters into numbers in order to evaluate an expression.

Subtract Minus, take away; the second number is the one being taken away.

Subtraction The process of taking away one number from another.

Subtrahend The number which is being subtracted from the larger number (minuend), e.g. 24 – 15 = 9

Sum (1) The answer to an addition.

Sum (2) A word sometimes used loosely to describe any calculation; 'doing sums' can involve any of the four operations, addition, subtraction, multiplication, division.

Summand One of the numbers being added together to make the sum.

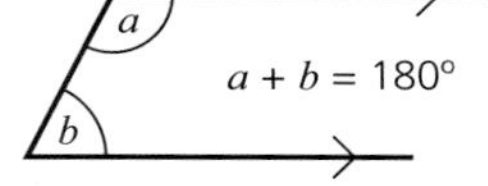

Supplementary angles Two angles which add up to 180 degrees.

Surd An expression involving roots which cannot otherwise be written down using whole numbers, fractions or terminating decimals; example $3\sqrt{2}$

Surface The outside of a solid shape.

Surface area The total area of each surface of a solid shape.

Survey A data collection exercise which usually involves a questionnaire of some kind.

Symbol Any sign used in mathematics to give a particular meaning, such as the symbol used for pi (π) or 'less than' (<).

Symmetrical Having symmetry of any kind.

Symmetry Sameness, especially with reflection or rotation.

Symmetry lines Mirror lines on a diagram which has reflection symmetry.

Systematic Having an organised approach, especially with regard to the creation of lists.

Table Any two-way chart which helps to organise the information it contains.

Tables Multiplication tables from 1×1 to 12×12

Tally Counting via the 'five bar gate' method or using ticks, which are later totalled to give the frequency.

Tally chart A diagram used to collect data by making marks in the appropriate row or column.

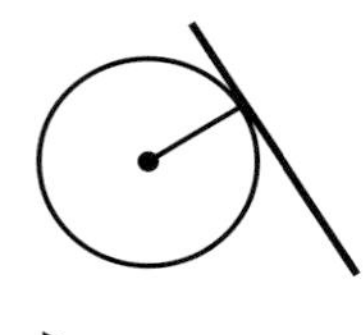

Tangent (1) Any straight line which just touches the circumference of a circle; it touches in one place and is always at right angles to the radius at that point.

Tangent (2) (tan) In a right-angled triangle, the tangent of an angle is equal to the ratio opposite : adjacent or opposite ÷ adjacent.

Tax Any form of duty added on to the price of something.

Temperature Measured by a thermometer, usually in degrees Celsius.

Ten, tens The place value position to the left of the Units column. H **T** U.t h

Tenth, tenths (1) The place value position to the right of the decimal point. H T U.**t** h

Tenths (2) The result when a whole is divided into ten equal pieces.

Term (1) A sequence made of a list of numbers generated by a rule; each number in the list is called a term.

Term (2)
A component of an algebraic expression, separated by + or – signs, e.g. 'like terms' or 'x^2 terms'.

Terminating decimal
A fraction converted into decimal form which terminates (stops), e.g. $\frac{3}{4} = 0.75$

Tessellate
Fit together on a flat surface so that no gaps are left; rectangles tessellate but circles do not.

Tessellation
Shapes fitted together to cover a plane without gaps.

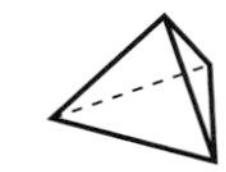

Tetrahedron
A regular triangular-based pyramid.

Theoretical probability
Assuming all outcomes are equally likely, the theoretical probability is (the number of ways of getting what you want) ÷ (the total number of equally likely outcomes).

Theory
Any idea which has support or evidence, but needs proof before it is accepted as true fact.

Therefore
As a result; following on from which; my conclusion is, etc.

Thermometer
An instrument used to measure temperature.

Thickness
The measurement of height for solid objects which are almost flat.

Thirds
The result when a whole is divided into three equal pieces.

Thousand, thousands
The place value position to the left of the Hundreds column. **Th** H T U.t h th

Thousandth, thousandths
The place value position to the right of the Hundredths column. Th H T U.t h **th**

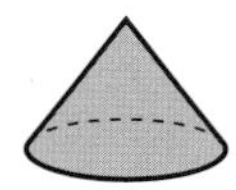

Three-dimensional (3D)
Solid, such that it may be picked up or viewed from different positions.

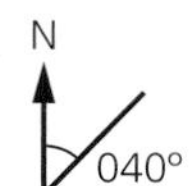

Three-figure bearing
A direction given as degrees clockwise from north.

Through 90°, etc.
A rotation needs a rotation centre to determine the point of turning, and an angle of rotation, e.g. 'through 180 degrees', to say through how much the object is to be turned.

Time
Any interval along the line from past, through present, to future, or any fixed position on that line.

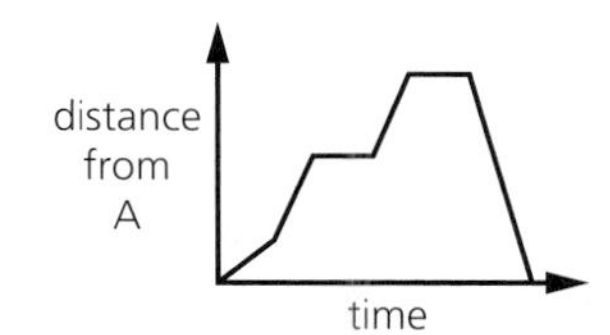

Time graph (line graph)
Any type of graph which shows the way something is changing over time, e.g. a temperature graph, a travel graph, a happiness graph.

Times — Multiply, e.g. as in 'three times four'; see also **Multiply**.

Timetable — A schedule of times, usually of events through the day or week, e.g. bus, train, and school lessons.

Title — A descriptive heading, especially important at the top of any graph, diagram, or new piece of work.

To one decimal place — A form of rounding which is accurate to the nearest tenth (to 1 d.p.). 0.36 → 0.4 to 1 d.p.

To the power of n — Repeatedly multiplied by itself n times ($n = 2$ for squaring and $n = 3$ for cubing).

Ton — An imperial unit for weighing very heavy objects; one ton is 160 stone (about 12 men) which is 2240 lb.

Tonne — One thousand kilograms, used for measuring the mass of very heavy objects; a metric tonne is only slightly lighter than an imperial ton.

Total — The result of adding; see also **Sum**.

Transform — Change, especially shape, for example by reflection, rotation, enlargement.

Transformation — A change, especially on a shape, for example, reflection, rotation, enlargement.

Translate — To slide – to transform only by a change of position.

Translation — The transformation by which a shape is translated.

Trapezium — A quadrilateral with one pair of parallel sides.

Trial and improvement — A method for solving equations by improving on each earlier trial.

Triangle — A plane shape with three sides.

Triangular number — One of the numbers which can be drawn as an equilateral triangle of dots; a number in the sequence 1, 3, 6, 10, 15, ... The nth triangular number is given by $\frac{1}{2}n(n+1)$.

Triangular prism — A solid shape with a continuous triangular cross-section which remains congruent between the two ends.

Trigonometry — The branch of mathematics dealing with the measurement of triangles.

True — Not false; correct, straight, honest.

Two-dimensional (2D) — Flat, such that it cannot be picked up; like a shadow or picture.

Two-way table — Used for sorting into rows and columns, e.g. a bus timetable, football results sheet.

Uncertain Having a probability which is less than one, and commonly used to mean with a probability less than 0.5

Unfair Biased; not favouring both sides equally.

Unit fraction Any fraction with a numerator of one, e.g. $\frac{1}{2}, \frac{1}{3}, \frac{1}{5}, \frac{1}{23}$

Unitary method Calculating with ratios in which you find out the value of one item before multiplying by the number required.

Units (1) The place value position to the left of the decimal point. H T **U**.t h

Units (2) The quantity being measured, e.g. cm, kg, min.

Universal set The greatest set of all; contains everything in the question ('the universe').

Unknown Anything not known, especially the variable letter in an equation.

Unlikely Having a low chance or likelihood.

Value The arithmetical result of substituting given values into an algebraic expression and evaluating the expression which remains.

Value added tax (VAT) A legally-required tax added on to the price of certain items.

Variable The letter in an equation representing the value to be found; the letters used in a formula or algebraic expression. e.g. $y = x + 3 \quad P = 2(l + w)$

Venn diagram A diagram using overlapping circles to show how two or more sets are related.

Verify Check that something is true.

Vertex A corner of a shape (pl. vertices).

Vertical Perpendicular to the horizontal; straight up.

Vertically opposite angles In an X formed by two lines crossing, angles facing each other rather than side by side are 'vertically opposite'.

Vertices Corners of a plane or solid shape (sing. vertex).

View The picture you get of a three-dimensional shape when seen from the side (elevation view) or the top (plan view).

Volume The amount of space inside a solid shape, measured in cubic cm (cm^3), cubic m (m^3), etc.

Volume factor The old volume was multiplied by this to obtain the new volume; the volume factor is the cube of the scale factor.

Vulgar fraction An ordinary fraction (proper fraction) with the numerator smaller than the denominator, e.g. $\frac{3}{5}$

Week A period of time lasting seven days.

Weight The force due to gravity exerted by something with mass; if there is no gravity, the mass stays the same but the weight becomes zero.

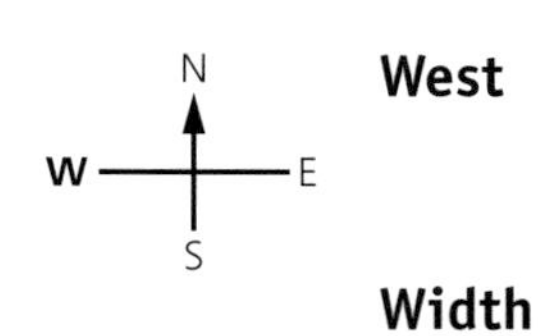

West Towards the setting sun.

Width A measure of how wide something is.

Word formula A set of instructions or a flow diagram, which tells you what to do with the input.

***x*-axis** The horizontal axis on a co-ordinate grid.

***x*-co-ordinate** The first number of a co-ordinate pair, to indicate the distance to the right of the origin (negative means to the left of the origin), e.g. (**3**, 2).

Yard A unit of length consisting of three feet or 36 inches (about 91.4 cm).

***y*-axis** The vertical axis on a co-ordinate grid.

***y*-co-ordinate** The second number of a co-ordinate pair, to indicate the distance above the x-axis (negative means below the x-axis), e.g. (3, **2**).

Year A period of time lasting 12 months; the time taken for Earth to travel once around the sun.

Zero (1) (place holder) e.g. the zero in 3028 means 'no hundreds'; the zero in 4.05 means 'no tenths'.

Zero (2) The reference point on a measuring scale; when measuring a length or an angle, make sure you line up with zero on the ruler/protractor and not the edge of the plastic.

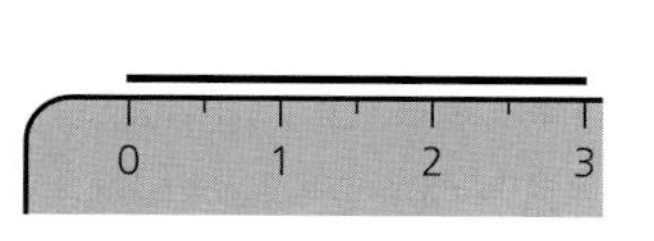

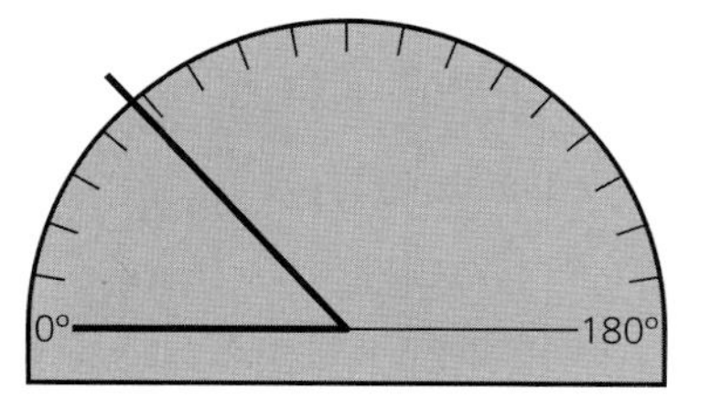

Zero (3) Nought; nothing – as in 'zero points' in the Eurovision Song Contest.

Zero (4) The last number in a countdown. e.g. 'five, four, three, two, one, zero'.

Exam-style questions answers

Chapter 1 (page 1)

1.1	(a)	15, 36, 45, 114, 1011	(2)
	(b)	12, 16, 24, 36, 52	(2)
	(c)	all of them!	(2)
1.2		2, 3, 5, 13, 17	(2)
1.3	(a)	6, 12 or 18	(1)
	(b)	3 or 6	(1)
	(c)	4	(1)
	(d)	3, 7	(2)
	(e)	12	(1)
	(f)	3, 7	(1)
	(g)	6, 7, 12 or 3, 4, 18	(2)
1.4	(a)	40, 45	(1)
	(b)	42, 39	(1)
	(c)	$^{-}3$, $^{-}7$	(2)
1.5	(a)	21, 28	(2)
	(b)	16, 26	(2)
	(c)	7, 3.5	(2)
1.6	(a)	$^{-}5$ m	(1)
	(b)	$^{-}0.9$ m	(2)
1.7	(a)	$^{-}7$ °C	(1)
	(b)	4 degrees	(1)
1.8	(a)	112	(1)
	(b)	1870	(1)
	(c)	2014	(1)
1.9	(a)	LVII	(1)
	(b)	CDXXIII	(1)
	(c)	MDCCLXXXIX	(1)
1.10	(a)	XXVIII	(1)
	(b)	twenty-eight	(1)
	(c)	28	(1)

(d) because 2Ds are the same as an M, etc. (1)

(e)

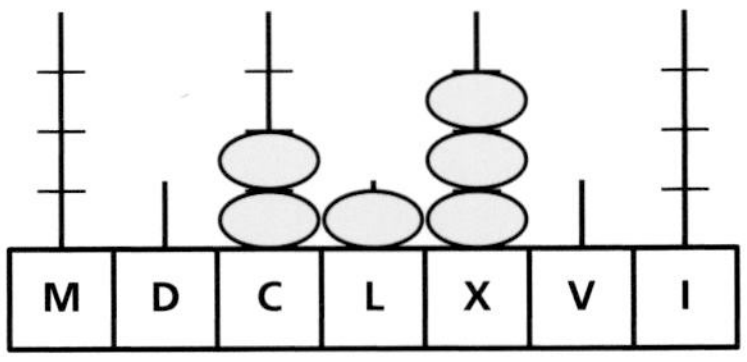

M D C L X V I

CCLXXX (280) (1)

1.11 (a) 309 (1)

(b) One hundred and forty-seven (1)

1.12 (a) 7 (7 units) (1)

(b) 300 (3 hundreds) (1)

(c) 100 times greater (7 tens as opposed to 7 tenths) (2)

1.13 21.3 5.30 78.35 (3)

1.14 (a) 3070 (1)

(b) 3.07 (1)

1.15 (a) 100 times (1)

(b) 1000 times (1)

1.16 (a) 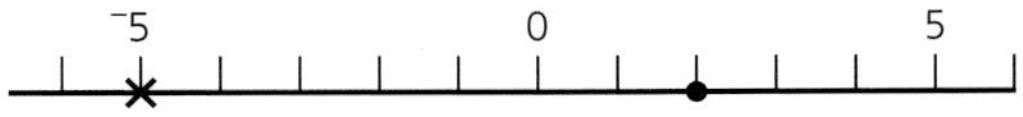 $^{-}5$ 0 5 (1)

(b) 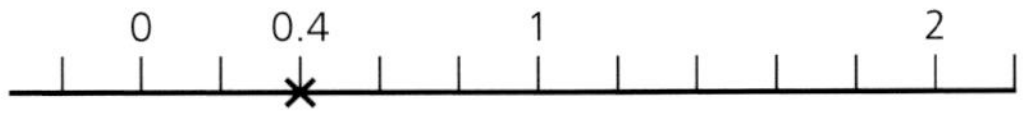 0 0.4 1 2 (1)

(c)  0 $\frac{3}{4}$ 1 2 3 (1)

1.17 (a) 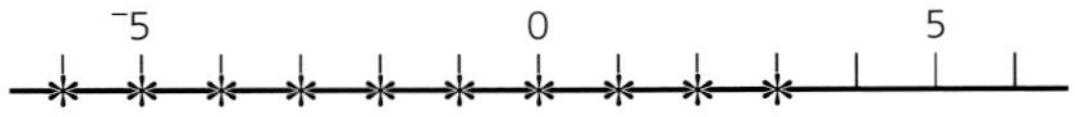 $^{-}5$ 0 5 (2)

(b) 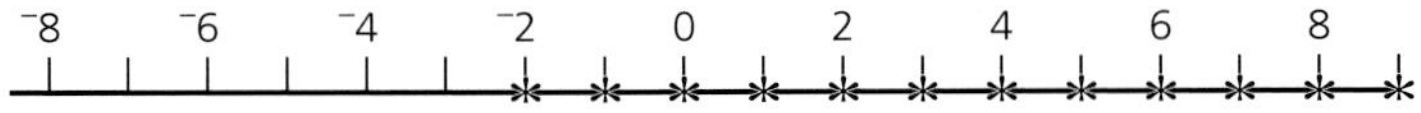 $^{-}8$ $^{-}6$ $^{-}4$ $^{-}2$ 0 2 4 6 8 (2)

1.18 (a) 132 213 231 312 321 (2)

(b) 8.1 7.91 7.36 7.3 6.37 (2)

1.19 (a)  0 5 10 (2)

(b) 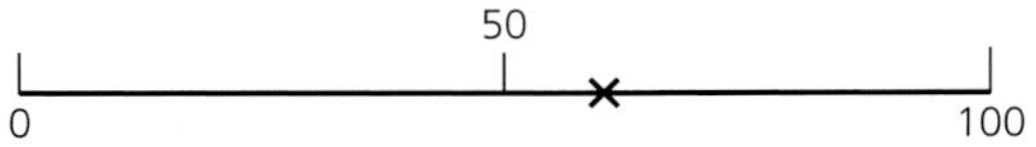 0 50 100 (2)

1.20 (a) 270 (1)

(b) 800 (1)

(c) 10 (1)

1.21 (a) 7000 (1)

(b) 6000 (1)

1.22 10 hours (3)

1.23 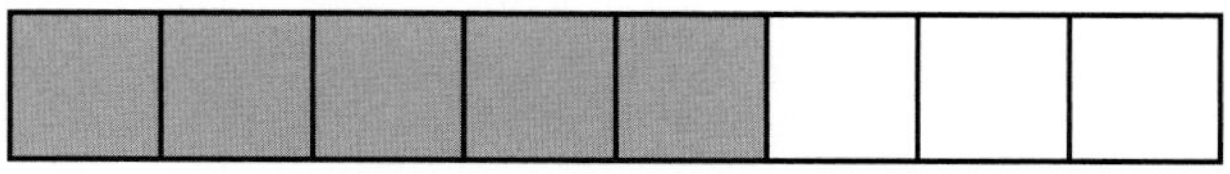 (2)

1.24 42 cm (2)

1.25 75 mm (2)

1.26 £30 (2)

1.27 (a) $\frac{1}{6}$ (1)

(b) $\frac{3}{4}$ (1)

1.28 (a) 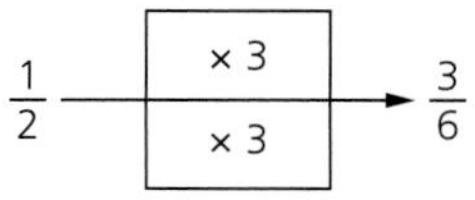

(2)

(b) 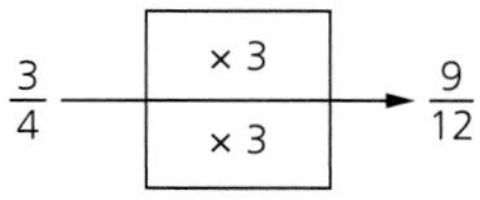

(2)

1.29 (9)

Fraction (in simplest form)	$\frac{1}{2}$	$\frac{3}{4}$	$\frac{2}{5}$	$\frac{7}{10}$	$\frac{3}{10}$
Decimal	0.5	0.75	0.4	0.7	0.3
Percentage	50%	75%	40%	70%	30%

1.30 (a) 5 : 2 (2)

(b) 2 : 7 (2)

Chapter 2 (page 17)

2.1 (a) 42 (1)

(b) 16 (1)

(c) 72 (1)

(d) 3 (1)

(e) 64 (1)

(f) 4 (1)

(g) 64 (1)

(h) 2 (1)

(i) 81 (1)

2.2 (a) 39 (1)

(b) 1 or 3 (1)

(c) 56 (1)

(d) 3 (1)

(e) 1 (1)

(f) 3 (1)

2.3 (a) 31, 37, 41, 43, 47 (3)

(b) 120 (1)

(c) $2 \times 3^2 \times 5 \times 31$ (3)

2.4 (a) $^-3$ $^-2$ $^-1$ 0 3 5 7 (2)

(b) (i) 2 (1)

(ii) $^-3$ (1)

(iii) $^-3$ (1)

(iv) 7 (1)

(c) (i) $^-10$ (1)

(ii) $^-20$ (1)

(iii) 12 (1)

(iv) $^-2$ (1)

(v) $^-\frac{1}{2}$ (1)

(vi) 3 (1)

2.5 (a) (i) 50.7 (1)

(ii) 0.0507 (1)

(iii) 507 (1)

(b) (i) 110.7 (1)

(ii) 11.07 (1)

(iii) 1.107 (2)

(iv) 820 (2)

2.6 (a) 1.32 2.13 2.31 3.12 3.21 (2)

(b) (i) $<$ (1)

(ii) $>$ (1)

(iii) $\geq$ (1)

(iv) $\leq$ (1)

2.7 (a) (i) 500 (1)

(ii) 400 (1)

(b) (i) 81.5 (1)

(ii) 81.5 (1)

2.8 (a) 40 (1)

(b) 9 (1)

2.9 (a) $\frac{5}{8}$ (1)

(b) $\frac{2}{5}$ $\left(\frac{8}{20} = \frac{4}{10} = \frac{2}{5}\right)$ (1)

(c) In $\frac{2}{3}$, 2 is the numerator and 3 is the denominator (1)

2.10 (a) (i) $\frac{6}{10}$ (1)

(ii) 0.6 (1)

(iii) 60% (1)

(b) $\frac{3}{5}$ (1)

2.11 (a) (i) $3\frac{2}{5}$ (1)

(ii) $\frac{12}{5}$ (1)

(b) (i) 1.75 (1)

(ii) 175% (1)

2.12 (a) $\frac{10}{13}$ (1)

(b) $1\frac{1}{6}$ (1)

(c) $2\frac{1}{12}$ (2)

(d) $\frac{1}{2}$ (1)

(e) $\frac{1}{6}$ (1)

(f) $\frac{7}{12}$ (2)

2.13 (a) $\frac{1}{2}$ (1)

(b) $\frac{5}{6}$ (2)

(c) $\frac{3}{4}$ (2)

(d) $\frac{16}{33}$ (3)

2.14 (a) (i) 6 : 11 (1)

(ii) 6 : 5 (1)

(iii) 5 : 6 (1)

(b) 2 : 3 (1)

2.15 To make 24 crunchies:

Cornflakes	600 g
Chocolate	500 g
Honey	40 ml
Sultanas	100 g

(4)

Chapter 3 (page 33)

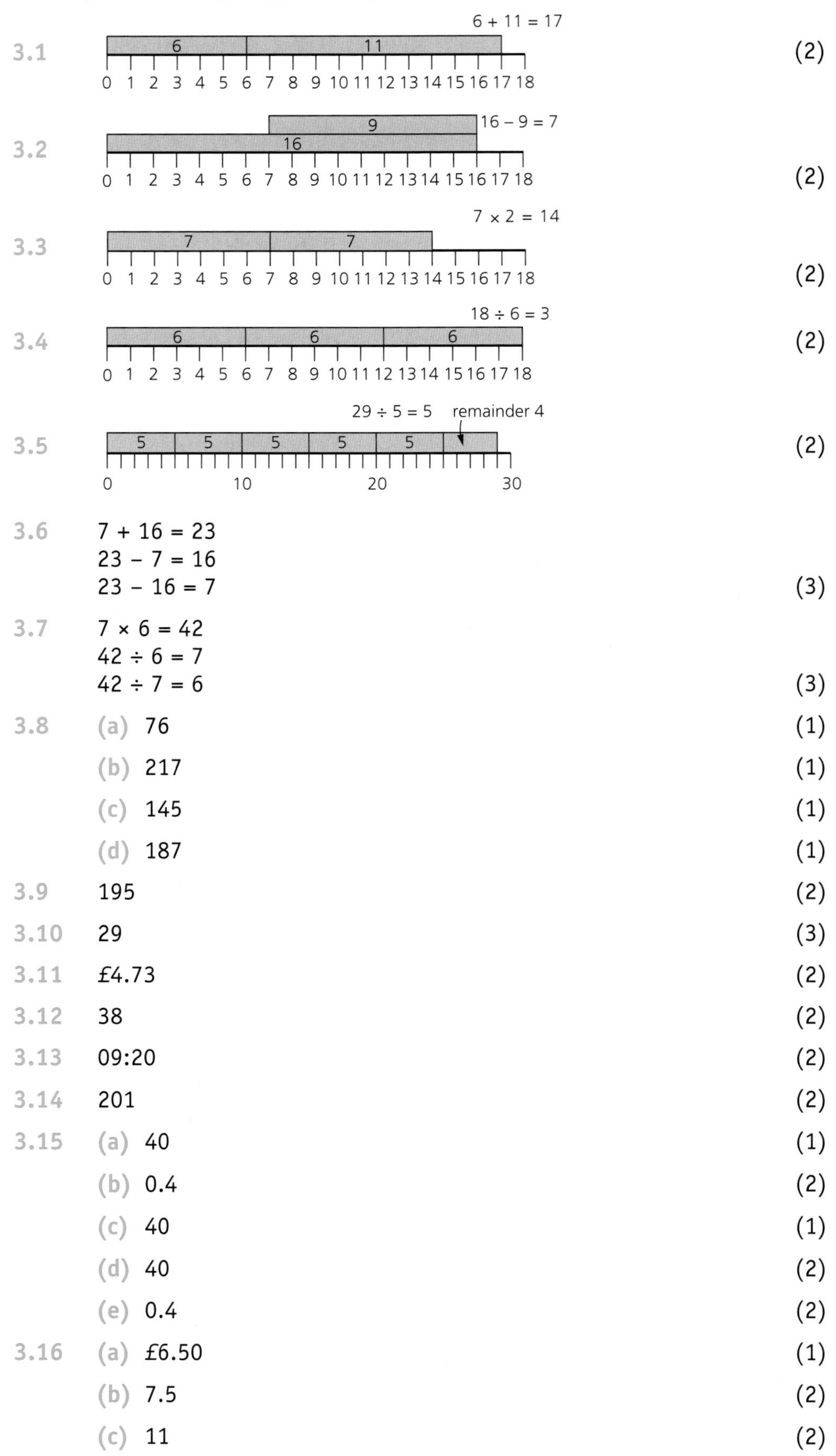

3.6 $7 + 16 = 23$
$23 - 7 = 16$
$23 - 16 = 7$ (3)

3.7 $7 \times 6 = 42$
$42 \div 6 = 7$
$42 \div 7 = 6$ (3)

3.8 (a) 76 (1)
(b) 217 (1)
(c) 145 (1)
(d) 187 (1)

3.9 195 (2)

3.10 29 (3)

3.11 £4.73 (2)

3.12 38 (2)

3.13 09:20 (2)

3.14 201 (2)

3.15 (a) 40 (1)
(b) 0.4 (2)
(c) 40 (1)
(d) 40 (2)
(e) 0.4 (2)

3.16 (a) £6.50 (1)
(b) 7.5 (2)
(c) 11 (2)

3.17 (a) 125 (1)

(b) 83 (2)

3.18 (a) 7.5 (1)

(b) 8.45 (2)

3.19 (a) 6.7 (1)

(b) 4.5 (2)

3.20 (a) 126 (2)

(b) 4056 (2)

3.21 (a) 241 (2)

(b) 143 (2)

3.22 (a) 24 (2)

(b) 2.6 (2)

3.23 (a) 71 remainder 6 (2)

(b) $71\frac{3}{7}$ (1)

3.24 (a) 46 (2)

(b) 80 (2)

3.25 (a) £16.50 (2)

(b) (i) 28 pence (2)

(ii) 4 pence (2)

Chapter 4 (page 42)

4.1 **(a)** 3100.1 (2)

(b) 2719.1 (2)

(c) 26 280 (2)

(d) 503 (2)

4.2 **(a)** 4.421 (2)

(b) 3.479 (2)

(c) 3.56 (2)

(d) 14.24 (2)

4.3 **(a)** 13 (1)

(b) 31 (1)

(c) 23 (2)

(d) $^{-}1$ (2)

4.4 **(a)** 1275 (1)

(b) 51 (2)

(c) 150 (2)

(d) 150 (2)

4.5	**(a)**	6		(3)
	(b)	1		(3)
4.6	**(a)**	£455.40		(3)
	(b)	£17.95		(3)
4.7	**(a)**	11.78		(1)
	(b)	3.871 794 872 ⟶ 3.87		(1)
	(c)	7.413 793 103 ⟶ 7.41		(1)
	(d)	12.378 553 44 ⟶ 12.4		(1)
4.8	**(a)**	195 364		(1)
	(b)	3.383 784 863 ⟶ 3.38		(1)
	(c)	1.181 665 75 ⟶ 1.18		(1)
	(d)	$2\frac{8}{11}$		(1)
4.9	**(a)**	105.683 176 9 ⟶ 106 cm^2		(1)
	(b)	3 760 000 000		(1)
4.10	**(a)**	£10.80		(1)
	(b)	15 min 12 s		(1)
	(c)	4 h 45 min		(1)
	(d)	4		(1)
	(e)	$\frac{2}{7}$		(1)
4.11	**(a)**	$1\frac{3}{5}$		(1)
	(b)	4 500 000 000		(1)
	(c)	0.005 07		(1)
	(d)	$1\frac{1}{3}$		(1)
	(e)	10		(1)
4.12	**(a)**	5.763 157 895		(2)
	(b)	5.76		(1)
	(c)	5.8		(1)
4.13	**(a)**	61.76		(1)
	(b)	60		(2)
	(c)	17.9		(2)
4.14	**(a)**	**(i)**	0.017 913 253 01	(2)
		(ii)	0.02	(1)
	(b)	**(i)**	⁻0.364 470 588 2	(2)
		(ii)	⁻0.364	(1)
4.15	**(a)**	**(i)**	$\frac{300}{30 + 10}$	(2)
		(ii)	$7\frac{1}{2}$	(1)

	(b) **(i)**	7.058 968 059	(1)
	(ii)	7	(1)
	(iii)	7.06	(1)
4.16	**(a)**	£40	(1)
	(b)	10	(1)
	(c)	10 kg	(1)
	(d)	£25	(1)
	(e)	£2.10	(1)
4.17	**(a)**	£550	(1)
	(b)	9 years	(1)
	(c)	1.5 m²	(1)
	(d)	64	(1)
	(e)	4000	(1)
4.18	**(a)**	£29	(2)
	(b)	£29.27	(2)
4.19	**(a)**	Megan is correct. Julian did think of 6	(3)

A suitable flow chart might look like this.

6 → + 3 → × 4 → − 6 → 30
30 → + 6 → 36 → ÷ 4 → 9 → − 3 → 6

(b) John is not correct. Moira thought of 106 (3)

106 → − 4 → ÷ 3 → + 11 → 45
45 → − 11 → 34 → × 3 → 102 → + 4 → 106

Chapter 5 (page 57)

5.1 Probable answers:

(a)	M	(1)
(b)	PP	(1)
(c)	PP	(1)
(d)	PP	(1)
(e)	MJ	(1)

5.2 Probable answers:

(a)	Measuring tape or metre rule	(2)
(b)	Measuring tape, or string and metre rule	(2)
(c)	Thermometer	(2)
(d)	Measuring jug or cylinder	(2)
(e)	Stopwatch	(2)

(f) Beam balance or scales (possibly find mass of a hundred and then divide by 100, depending on accuracy of scales) (2)

(g) Centimetre squared paper (probably draw round your foot and count squares) (2)

5.3 50 and 59 (2)

5.4 (a) 50p, 10p, 2p (1)

(b) 5p, 2p (2)

5.5 35 (3)

5.6 13 (the numbers are 4 and 9) (3)

5.7 (a) (i) A (1)

(ii) C (1)

(iii) B (1)

(b) For example: (4)

D

E

5.8 £1.50 each (John has £5.60 at the start; together they have £9.80 before buying the DVD) (3)

5.9 £6.05 (milk £1.29, bread £1.56, cheese £1.10; total £3.95) (3)

5.10 (a) $\frac{1}{5}$ (2)

(b) 54% (1)

(c) 11 g (2)

5.11 (a) 192 (2)

(b) (i) 1280 g or 1.28 kg (2)

(ii) 720 g (2)

(c) 520 g (2)

(d) 442 g (2)

(e) 6p (1)

(f) £4.80 (2)

Chapter 6 (page 65)

6.1 Probable answers:

(a) Try a simpler example first or try a practical approach. (1)

(b) Make an organised list. (1)

(c) Guess and check. (1)

(d) Look for a pattern. (1)

	(e)	Try a simpler example first.	(1)
	(f)	Find a formula and then test it before using it.	(1)

6.2 **(a)** (2)

Year	Number of flowers
1	3
2	5
3	7
4	9
5	11

	(b)	15 flowers	(2)
	(c)	Year 15	(2)
6.3	**(a)**	36	(2)
	(b)	**(i)** 10	(2)
		(ii) 16	(2)
		(iii) 8	(2)
		(iv) 2	(1)
6.4	**(a)**	28 or 91	(2)
	(b)	**(i)**	(2)
		(ii) Kite	(1)
		(iii) 1	(1)

6.5 **(a)** (2)

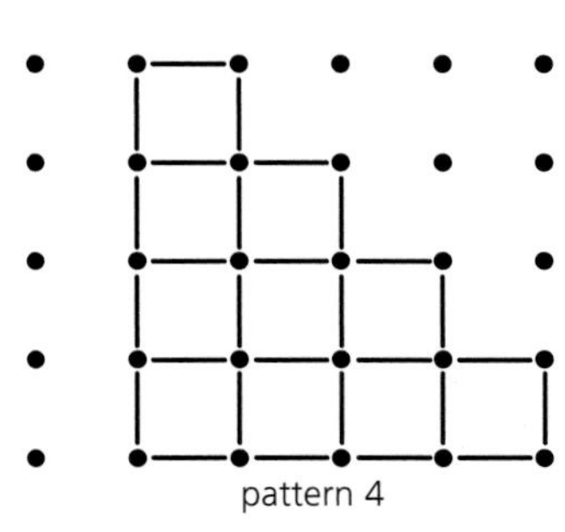

(b) (6)

Pattern number	Number of squares	Total number of dots	Perimeter (cm)
1	1	4	4
2	3	8	8
3	6	13	12
4	10	19	16
5	15	26	20

(c) (3)

6	21	34	24
7	28	43	28
8	36	53	32

For checking purposes only, each row's figures are given by:

n	$\frac{1}{2}n(n+1)$	$\frac{1}{2}(n^2+5n+2)$	$4n$

6.6 **(a)** £50 (2)

(b) 50 litres (2)

6.7 **(a)** £41.93 (2)

(b) **(i)** £27.95 (2)

(ii) £22.05 (2)

(c) £16.95 (division by factors recommended) (2)

6.8 **(a)** flour 12 ounces; margarine 9 ounces; sugar 6 ounces (3)

(b) She will use: flour 10 ounces; margarine $7\frac{1}{2}$ ounces; sugar 5 ounces.

She can bake at most 15 cakes ($\frac{5}{6}$ of 18). (2)

Chapter 7 (page 74)

7.1 (a) 6 (1)

(b) 23 (1)

(c) 3 (1)

(d) 1 (1)

(e) 9 (2)

7.2 (a) 9 years (1)

(b) 6 cm (1)

7.3 (a) 3 (1)

(b) $^{-}1$ (2)

(c) 10 (1)

7.4 (a) 9 (1)

(b) 23 (1)

(c) 6 (1)

(d) 54 (1)

7.5	**(a)**	7	(1)
	(b)	$^{-}2$	(2)
	(c)	2	(2)
7.6	**(a)**	2	(2)
	(b)	5	(2)
7.7			(2)
7.8	**(a)**	12, 14	(1)
	(b)	22, 26	(2)
	(c)	5, 0	(2)
	(d)	$1\frac{1}{2}, 1\frac{3}{4}$	(2)
7.9	**(a)**	10, 8	(2)
	(b)	16, 22	(2)
7.10	**(a)**	× 5 machine	(2)
	(b)	× 3 machine	(2)
7.11	**(a)**		(2)

input	+ 2	output
0	→	2
1	→	3
2	→	4
3	→	5

(b) (2)

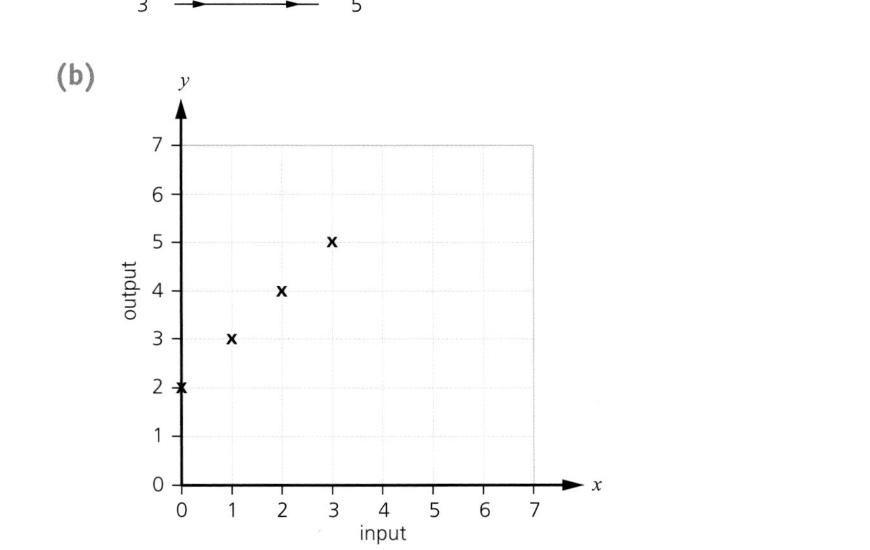

7.12	**(a)**	6	(1)
	(b)	8	(1)
	(c)	Output = input − 1	(2)

7.13

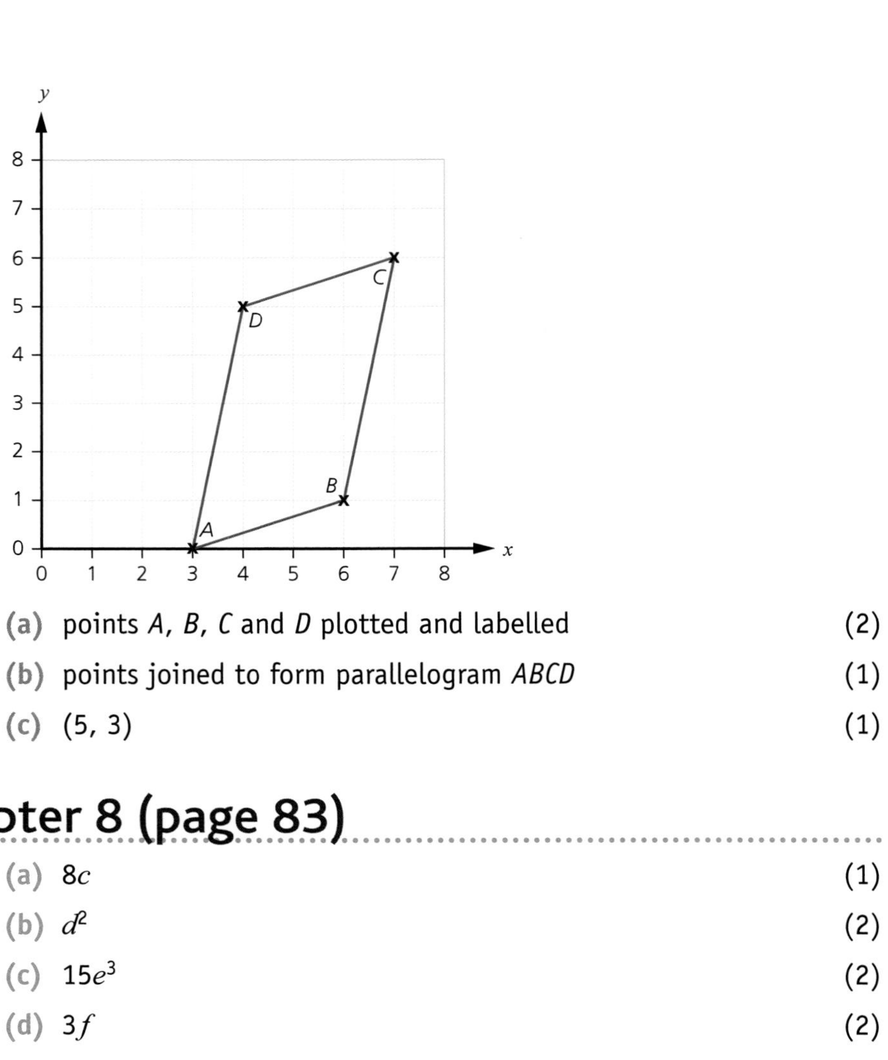

(a) points A, B, C and D plotted and labelled (2)

(b) points joined to form parallelogram $ABCD$ (1)

(c) (5, 3) (1)

Chapter 8 (page 83)

8.1 (a) $8c$ (1)

(b) d^2 (2)

(c) $15e^3$ (2)

(d) $3f$ (2)

(e) 2 (2)

8.2 (a) $3(h - 4)$ (2)

(b) $5(4e - 3f)$ (2)

(c) $2e(e + 2)$ (2)

(d) $4p(p + 2q)$ (2)

8.3 (a) $4a + 12b$ (1)

(b) $6q + 4r$ (2)

(c) $2m + 3n$ (3)

8.4 (a) $^{-}2$ (2)

(b) 120 (2)

(c) 2 (2)

(d) $^{-}9$ (3)

8.5 (a) 1 (1)

(b) $^{-}8$ (1)

(c) $^{-}18$ (2)

(d) 12 (2)

8.6 (a) $t + 3$ (1)

(b) $4t$ (1)

(c) $t - 5$ (1)

(d) $\frac{t}{2}$ (1)

(e) t^2 (1)

8.7 (a) $b + 2$ (1)

(b) $3b$ (1)

(c) $5b + 2$ (2)

(d) $5b + 2 = 22$, leading to $b = 4$ (2)

8.8 (a) $5x$ (1)

(b) $x + 3$ (1)

(c) $7(x + 3)$ or $7x + 21$ (2)

(d) $3x + 4(x + 3)$, leading to $7x + 12$ (2)

(e) $7x + 12 = 467$, leading to $x = 65$ (3)

(f) £7.34 (2)

8.9 (a) $q = 4$ (1)

(b) $p = 4$ (1)

(c) $n = 3$ (2)

(d) $y = 2$ (2)

(e) $x = {}^{-}7\frac{1}{2}$ (3)

8.10 (a) $s = 4$ (2)

(b) $a = \frac{1}{2}$ (2)

(c) $r = 1\frac{1}{2}$ (2)

(d) $y = 15$ (3)

(e) $m = 26$ (2)

8.11 (a) $a = 7$ (2)

(b) $u = 2$ (3)

(c) $v = {}^{-}1$ (3)

(d) $c = 28$ (2)

(e) $v = \frac{8}{3}$ or $2\frac{2}{3}$ (3)

(f) $s = 7$ (3)

8.12 6.57 (when x is 6, y is 24; when x is 7, y is 35 so it is reasonable to suppose that when y is 30 the value of x must be a little more than 6.5) (5)

8.13 (a) (i) $x < 2\frac{1}{5}$ (2)

(ii) 1, 2 (1)

(b) (i) $n \leq 6$ (3)

(ii) 6 (1)

8.14 (a) 31, 27 (2)

(b) 21, 31 (2)

(c) 50, 25 (2)

(d) 23, 28 (2)

(e) 256, 1024 (2)

(f) 1, 0 (2)

8.15 (a) 6, 16, 36, 76 (2)

(b) 'add 3' (1)

8.16 (a) 299 (1)

(b) $n = 334$ gives $T_n = 1001$ (2)

8.17 (2)

x	$^{-}1$	0	1	2	3	4
y	3	4	5	6	7	8

8.18 (2)

x	$^{-}2$	$^{-}1$	0	1	2	3
y	4	1	0	1	4	9

8.19

y
$x = {}^{-}3$
$x = 4$
$y = 5$
4
2
$^{-}4$ $^{-}2$ 0 2 4 6 x

(a) graph of $x = 4$ drawn (shown in green above) (1)

(b) graph of $y = 5$ drawn (shown in blue above) (1)

(c) graph of $x = {}^{-}3$ drawn (shown in red above) (1)

8.20

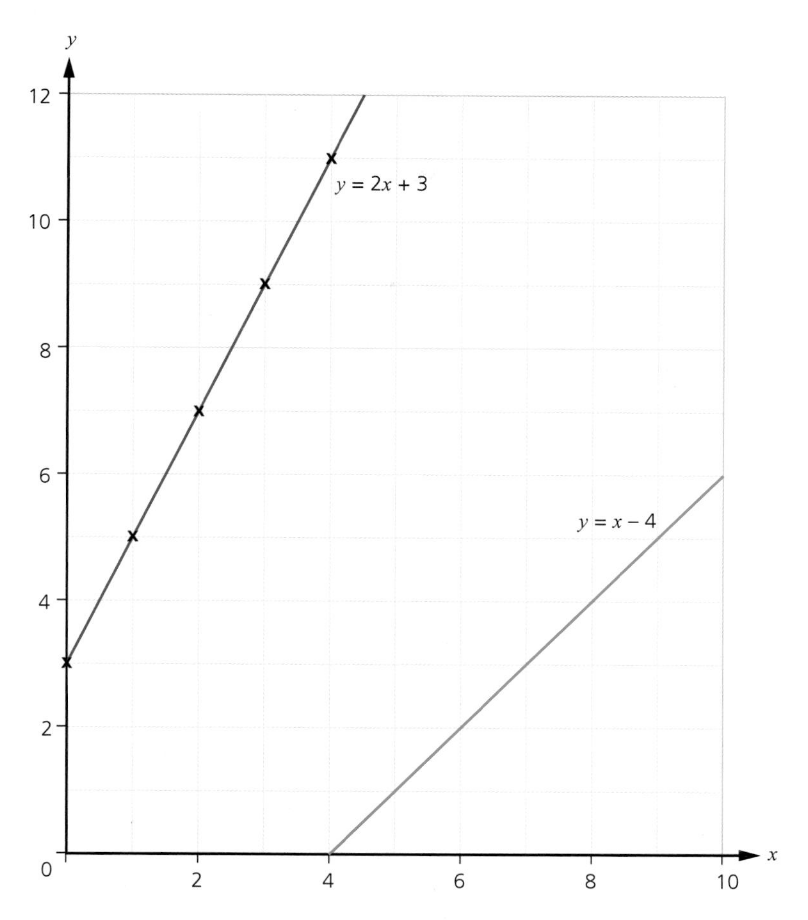

(a) points plotted and graph drawn of $y = 2x + 3$ (shown in red above) (2)

(b) graph drawn of $y = x - 4$ (shown in green above) (2)

(c) ($^-7$, $^-11$) (1)

8.21 (2)

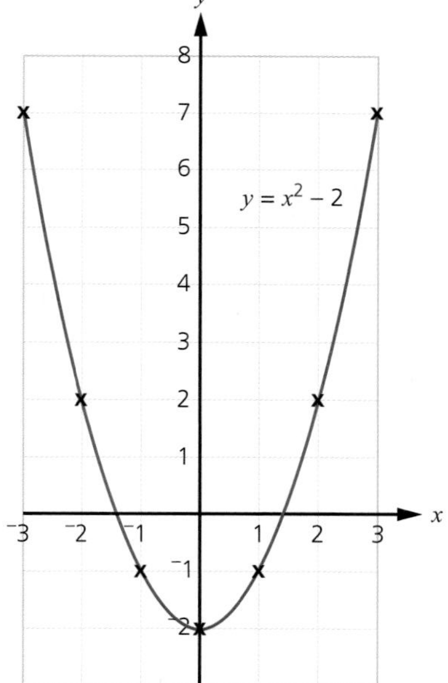

8.22 (a) (3)

x	$^-4$	$^-3$	$^-2$	$^-1$	0	1	2
x^2	16	9	4	1	0	1	4
$2x$	$^-8$	$^-6$	$^-4$	$^-2$	0	2	4
y	8	3	0	$^-1$	0	3	8

(b) (3)

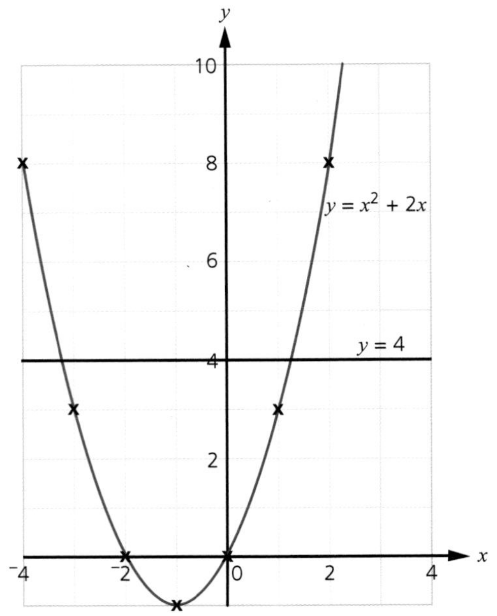

(c) $x = 1.2$ and $x = {}^{-}3.2$ (2)

(d) $x = {}^{-}1$ (1)

8.23 (a) (2, 4) (1)

(b) $4 = 2 + 2$ and $4 = 6 - 2$ (2)

8.24 (a) $2x - 3 = x + 1$ (1)

(b) $x = 4$ (2)

(c) $y = 5$ (1)

(d) check (1)

8.25 (a) $4y = 8$ (1)

(b) $y = 2$ (1)

(c) $x = 1$ (2)

8.26 (a) $3x + 2y = 110$ (1)

(b) $5x = 4y$ or $5x - 4y = 0$ (1)

(c) $x = 20$, $y = 25$ (4)

(d) 10 apples and 1 pear, 5 apples and 5 pears, just 9 pears (3)

Chapter 9 (page 103)

9.1 (a) 23 (2)

(b) 94 (2)

9.2 (a) 700 ml (2)

(b) 225 ml (2)

9.3 (a) (i) 1230 mm (1)

(ii) 1.23 m (1)

(b) (i) 40 000 g (1)

(ii) 88 pounds (40×2.2) (2)

9.4	(a) 22 cm		(1)
	(b) 30 cm²		(1)
9.5	(a) 12 cm³		(2)
	(b) 36 cm³		(2)
9.6	(a) (i) 13:10		(1)
	(ii) 07:35		(1)
	(b) (i) 2.30 p.m.		(1)
	(ii) 9.45 p.m.		(1)
9.7	(a)		(3)

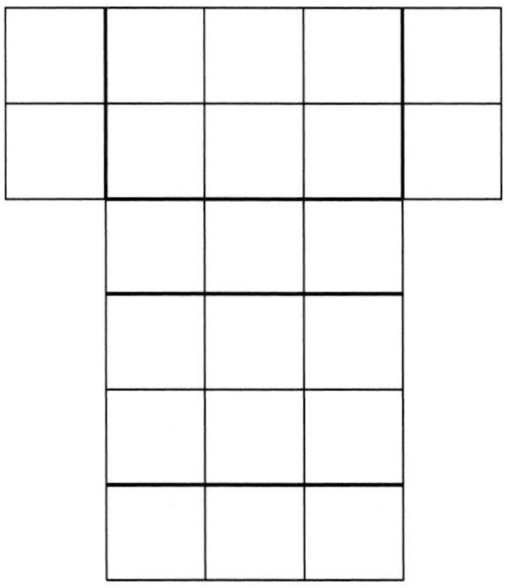

	(b) 22 cm²		(3)
9.8	(a) A rhombus, B isosceles trapezium		(2)
	(b)		(3)

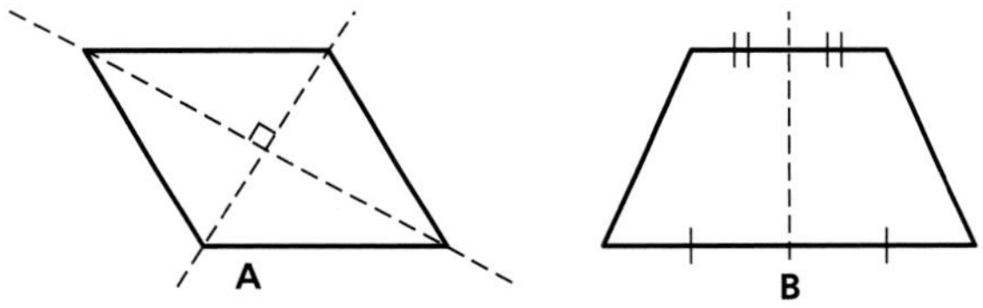

A B

	(c) A rotational symmetry order 2, B no rotational symmetry (sometimes called 'order 1')		(2)
9.9	B, D and E		(4)
9.10	(a)		(3)

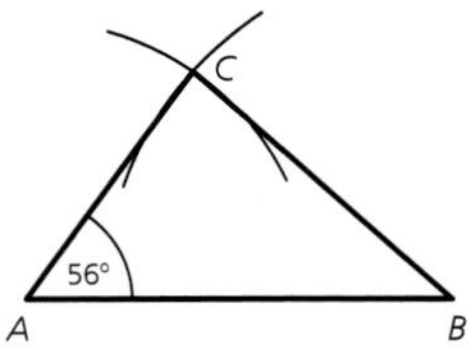

C 56° A B

	(b) angle *BAC* 56°		(1)
9.11	(a) (i) obtuse		(1)
	(ii) acute		(1)
	(b) scalene		(1)
9.12	(a) N		(1)
	(b) S		(1)
	(c) NE		(1)
	(d) SE		(1)

9.13 (3)

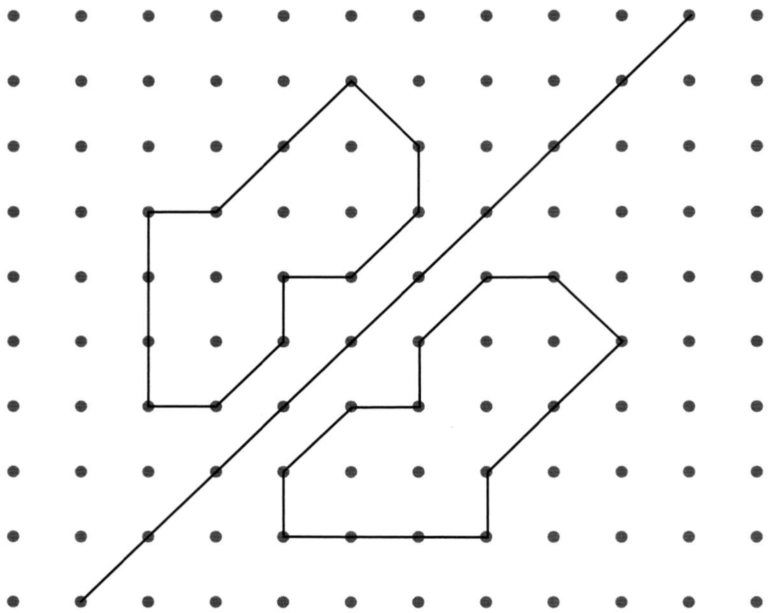

9.14

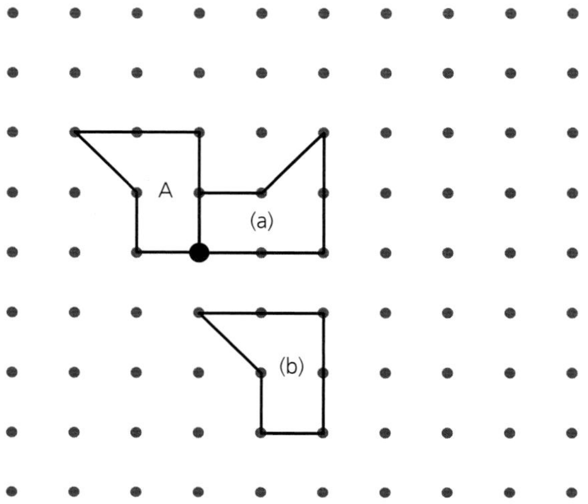

A

(a)

(b)

(a) rotation (2)

(b) translation (2)

Chapter 10 (page 121)

10.1 (a) (i) 130 mm (1)

(ii) 0.13 m (1)

(iii) 5 inches (1)

(b) 1.8 m (1)

(c) 960 km (2)

10.2 (a) 5 ha (1)

(b) (i) $100\,cm^2$ (1)

(ii) $1000\,cm^3$ (1)

10.3 (a) (i) 12 cm/minute (1)

(ii) 0.2 cm/s (1)

(iii) 720 cm (7.2 m) (1)

(b) (i) 10.5 miles (2)

(ii) 09:20 (2)

10.4 (a) (i) (4)

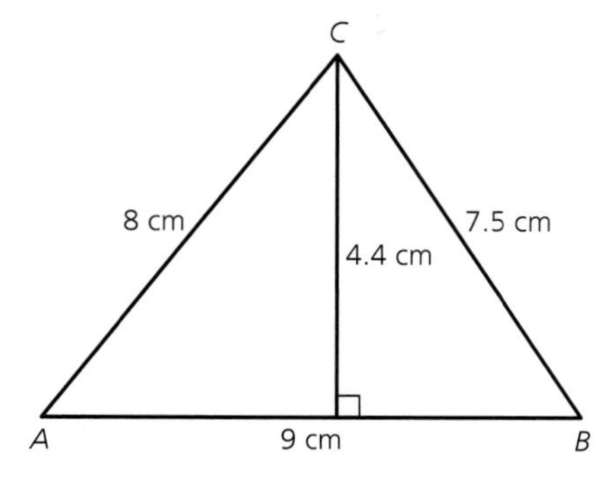

(ii) 6.2 cm (2)

(b) (i) (3)

7.1 cm

(ii) 28.4 cm (2)

10.5 (a) (2)

4 cm
7 cm
5 cm

(b) 140 cm^3 (2)

(c) 166 cm^2 (4)

10.6 (a) 36 cm (3)

(b) 8 mm (0.8 cm) (1)

10.7 (a) about 30 cm (2)

(b) about 75 cm^2 (2)

10.8 (a) 37.7 cm (2)

(b) 113 cm^2 (3)

10.9 (a) 100 cm^2 (1)

(b) 157 cm^2 (3)

(c) 257 cm^2 (1)

(d) 62.8 cm (2)

10.10 (a) (i) 3.5 cm (3)

(ii) 38.5 cm^2 (2)

(b) 424 cm^3 (2)

(c) 2.5 cm (3)

10.11 (a) Equilateral triangle, parallelogram, regular pentagon (3)

(b) (3)

rhombus (i)

kite (ii)

isosceles trapezium (iii)

(c) (3)

square (i)

kite (ii)

hexagon (iii)

10.12 (a) regular hexagon – 6 lines of symmetry, rotation symmetry order 6 (2)

(b) parallelogram – no lines of symmetry, rotation symmetry order 2 (1)

(c) isosceles triangle – one line of symmetry, no rotation symmetry (or rotation symmetry order 1) (1)

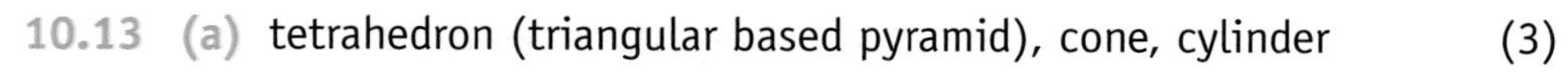

10.13 (a) tetrahedron (triangular based pyramid), cone, cylinder (3)

(b) (1)

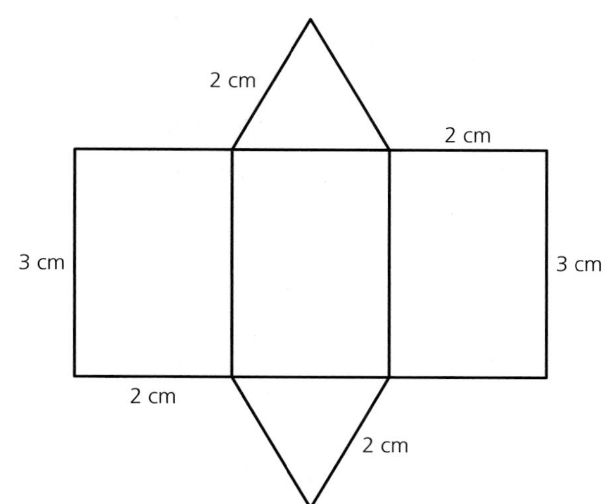

10.14 (a) square, rhombus (2)

(b) square, rectangle, rhombus, parallelogram (2)

(c) square, rhombus, kite (2)

(d) square, rectangle, isosceles trapezium (2)

(e) parallelogram (2)

(f) kite, isosceles trapezium (2)

10.15 (a) 120° (1)

(b) (i) 24° (1)

(ii) 156° (1)

(iii) 2340° (2)

10.16

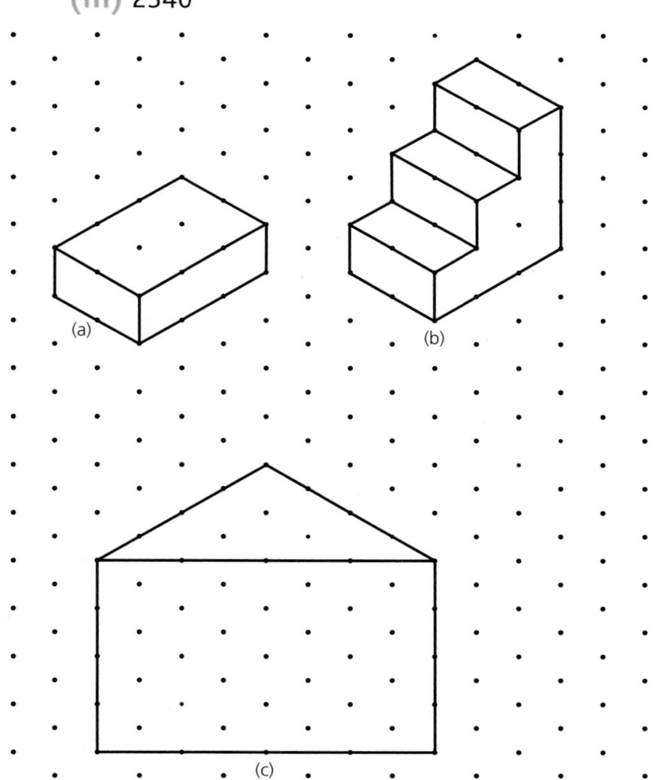

(a) cuboid drawn (1)

(b) step structure drawn (3)

(c) triangular prism drawn (3)

10.17 (a) A is obtuse, B is acute, C is reflex (3)

(b) (3)

10.18 (a) g (1)

(b) d (1)

(c) g (1)

10.19 (a) $a = 110°$, $b = 110°$ (2)

(b) $c = 35°$, $d = 75°$ (4)

10.20 (a) $p = 135°$, $q = 135°$ (2)

(b) $r = 22.5°$, $s = 247.5°$ (4)

10.21 (a) (2)

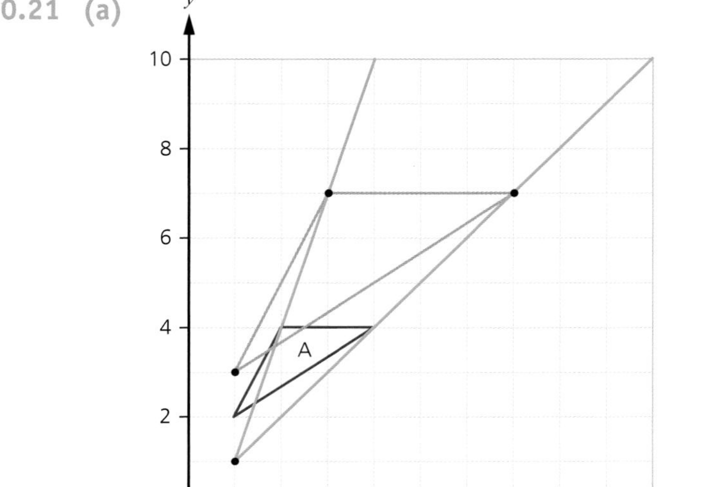

(b) 4 times (1)

10.22 (a) (4)

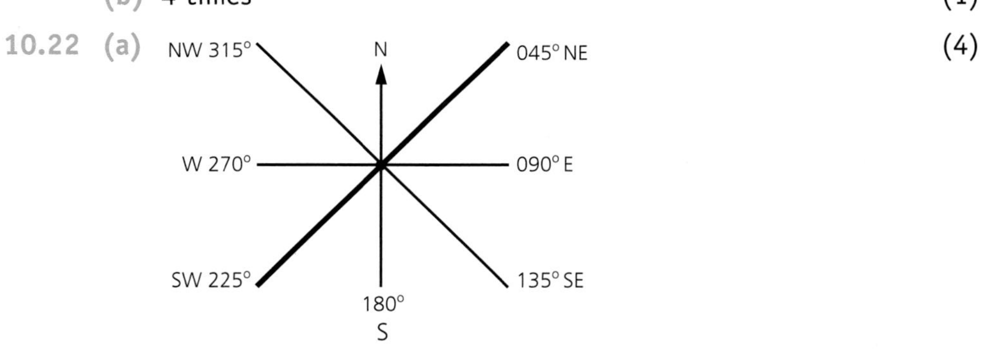

(b) SE, 135° (2)

10.23 (a)

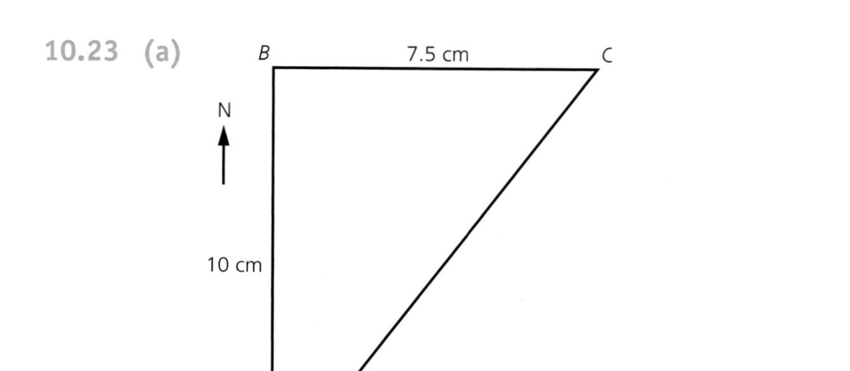

(3)

(b) (i) 037° (1)

(ii) 217° (2)

(iii) 25 m (2)

10.24 (a) 6.40 cm (2)

(b) 7.21 cm (2)

(c) 5.66 cm (3)

10.25

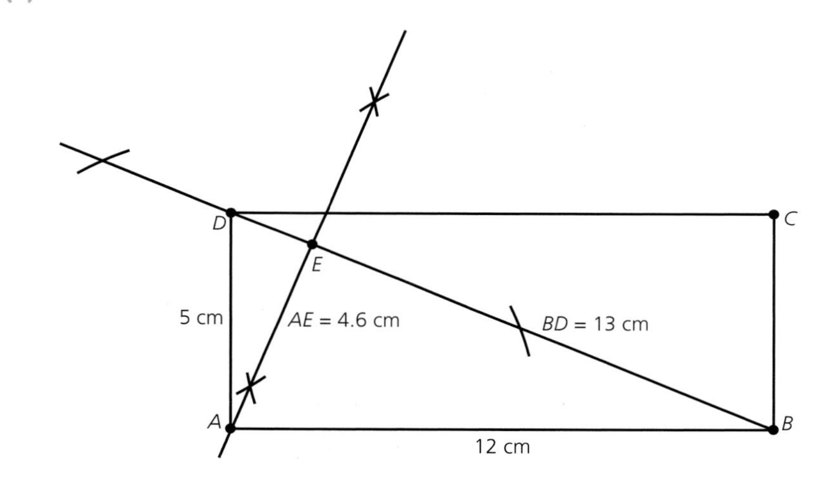

(a) 13 cm, verified by Pythagoras' theorem (2)

(b) (Construction lines shown, point *E* marked) (2)

(c) between 4.6 and 4.7 cm (1)

(d) about 30 cm^2; triangle *ABD* is exactly half of rectangle *ABCD*, so must be half the area. (2)

Chapter 11 (page 144)

11.1 **(a)** (4)

	odd	not odd
not prime	1 9	4 6 8
prime	3 5 7	2

(b) (4)

odd: 1 9
odd and prime: 3 5 7
prime: 2
outside: 4 6 8

11.2 **(a)** 7 (1)

(b) 36 (2)

11.3 **(a)** 17 (1)

(b) 120 (3)

(c) 5 (1)

(d) $\frac{1}{3}$ (2)

11.4 **(a)** 6 (1)

(b) 2 (1)

(c) 1 (1)

(d) 3 (2)

11.5 38 kg (3)

11.6 **(a)** (4)

Height (m)	Tally marks	Frequency
1.25 to 1.29	I	1
1.30 to 1.34	II	2
1.35 to 1.39	II	2
1.40 to 1.44	卌	5
1.45 to 1.49	卌 I	6
1.50 to 1.54	III	3
1.55 to 1.59	I	1
	Total	20

(b) (4)

frequency: 0, 2, 4, 6
1.25–1.29 1.30–1.34 1.35–1.39 1.40–1.44 1.45–1.49 1.50–1.54 1.55–1.59
height (m)

(c) 1.45 to 1.49 (1)

11.7 (a) 09:30 (1)

(b) 20 litres (1)

(c) 12:30 (1)

(d) 30 litres (2)

11.8 (a) (i) 2 (1)

(ii) 3 (1)

(b) H 1 H 2 H 3 T 1 T 2 T 3 (3)

11.9 (a) (i) (2)

Score 5	
Red	Blue
4	1
3	2
2	3
1	4

(ii) (2)

Score 8	
Red	Blue
6	2
5	3
4	4
3	5
2	6

(b) Minnie is not right. There is an equal chance of rolling two sixes as rolling two of any number. (1)

11.10 (a) even chance (1)

(b) impossible (1)

(c) unlikely (1)

(d) likely (1)

(e) even chance (2)

11.11 (a) impossible certain (1)

(b) impossible certain (1)

(c) impossible certain (1)

Chapter 12 (page 156)

12.1 (a) (2)

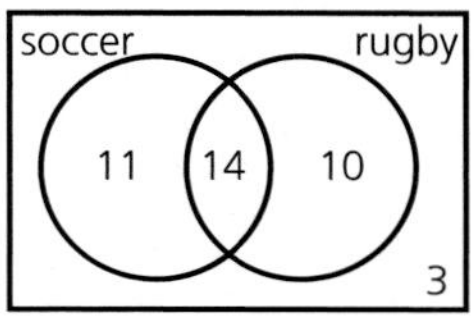

soccer rugby 11 14 10 3

	soccer	not soccer
not rugby	11	3
rugby	14	10

(b) 21 (1)

12.2 (a) $\frac{2}{5}$ (2)

(b) 35% (2)

(c) 90° (1)

12.3 **(a)** 12° (1)

(b) 3 (1)

(c) 1 (2)

(d) 20% (2)

12.4 **(a)** 60–69 (1)

(b) 21 (1)

12.5 **(a)** (3)

Score	Tally	Frequency
1	III	3
2	IIII	4
3	II	2
4	III	3
5	III	3
6	~~IIII~~	5
	Total	20

(b) (3)

(c) 6 (1)

(d) 3.7 (3)

(e) 4 (2)

12.6 **(a)** 25 (1)

(b) (3)

Number of peas	Tally	Frequency
20–22	III	3
23–25	~~IIII~~	5
26–28	~~IIII~~ II	7
29–31	III	3
	Total	18

(c) 26–28 (1)

12.7 **(a)** **(i)** 6 (1)

(ii) 3 (1)

(iii) $2\frac{1}{2}$ (2)

(b) Rovers had the larger range of scores. Wanderers scored in every match, were more consistent and scored the highest number of goals overall (35 compared to 29 scored by Rangers). (2)

12.8 **(a)** 362 kg (2)

(b) 280 kg (1)

(c) 41 kg (2)

12.9 **(a)** (4)

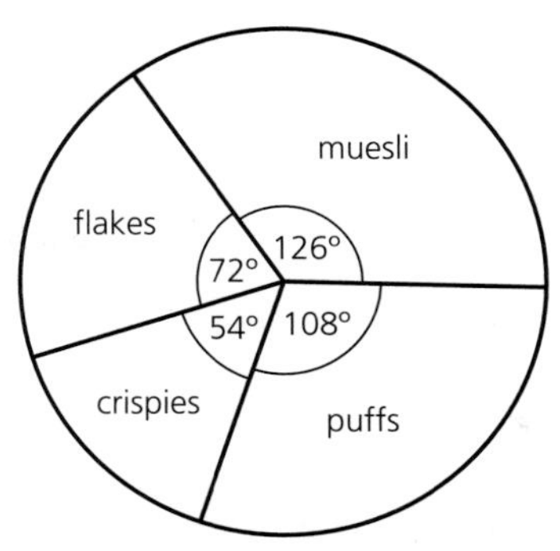

(b) **(i)** 9 (2)

(ii) 21 (2)

12.10 **(a)** (All answers are approximate.)

(i) \$56 (2)

(ii) £50 (2)

(b) **(i)** \$80 at the duty free shop (2)

(ii) £15 (2)

12.11 **(a)** (3)

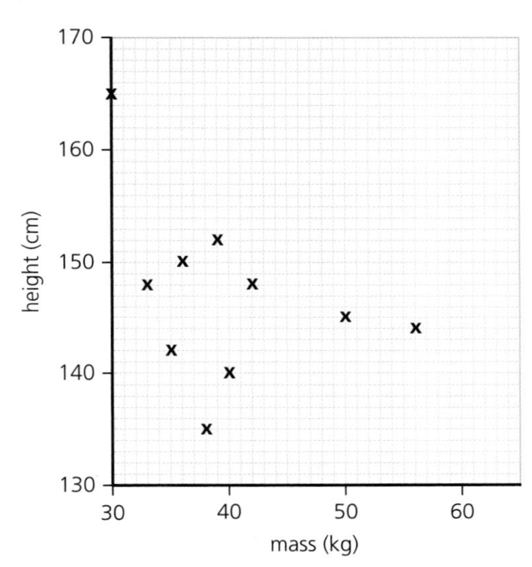

(b) It is not possible to draw a line of best fit from the data since there appears to be little or no correlation – the sample of girls includes one very tall girl with low mass, for example. (1)

(c) It is possible to make a guess, based upon the data – perhaps about 145 cm. (2)

Note: This question is not typical of scattergraph examination questions, but is included here to make you think!

12.12 **(a)** **(i)** $\frac{2}{5}$ (1)

(ii) (1)

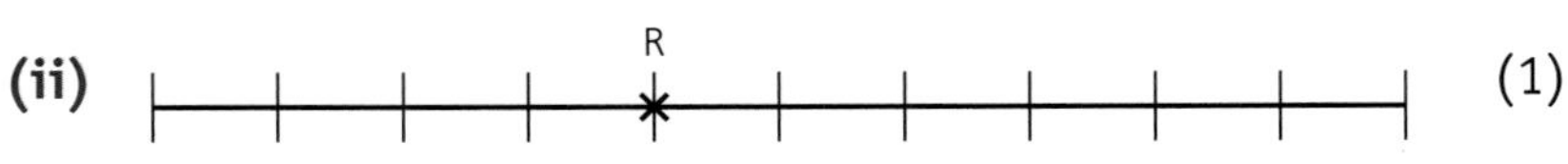

(b) $\frac{1}{3}$ (1)

12.13 (a) Yes, possibly – the spinner appears to be biased towards 6 (and 1 and 5 which are adjacent to 6). This could be because the spindle is slightly off centre (nearer to 3) or there could be a 'weight' under the 6

To be sure, it would be necessary to investigate with a larger number of spins. (1)

(b) About 30% (1)

12.14 (a) (2)

		Score on red die			
		1	2	3	4
Score on green die	1	2	3	4	5
	2	3	4	5	6
	3	4	5	6	7
	4	5	6	7	8

(b) (i) $\frac{3}{16}$ (1)

(ii) $\frac{13}{16}$ (1)

(c) $\frac{3}{4}$ (2)

12.15 (a) (3)

	C	L	A	R	E
M	MC	ML	MA	MR	ME
A	AC	AL	AA	AR	AE
R	RC	RL	RA	RR	RE
I	IC	IL	IA	IR	IE
E	EC	EL	EA	ER	EE

(b) $\frac{3}{25}$ (2)

(c) $\frac{6}{25}$ (2)

Test yourself answers

Chapter 1 (page 1)

1

1	2	3	4	5	6
7	8	9	10	11	12
13	14	15	16	17	18
19	20	21	22	23	24
25	26	27	28	29	30
31	32	33	34	35	36

(a) 3 and every third number circled

(b) 4 and every fourth number with a cross

(c) The numbers which are circled and have a cross are multiples of 12 (common multiples of 3 and 4).

2

×	2	3	4	5	6	7	8
2	4	6	8	10	12	14	16
3	6	9	12	15	18	21	24
4	8	12	16	20	24	28	32
5	10	15	20	25	30	35	40
6	12	18	24	30	36	42	48
7	14	21	28	35	42	49	56
8	16	24	32	40	48	56	64

(a) multiples of 9 shaded

(b) square numbers circled

(c) 12 and 24 (4 times each); 16 appears 3 times

3 (a) A prime number is a number with no factors except itself and 1

(b) 2, 3, 5, 7, 11, 13, 17, 19, 23, 29, 31, 37, 41, 43, ...

4 (a) (i) 345 354 435 453 534 543

(ii) 350 350 440 450 530 540

(iii) all of them! – the digits add to a multiple of 3

(iv) 345 and 435

(v) 54 300

(vi) 34.5

(b) (i) $\frac{3}{5}$

(ii) $\frac{4}{5}$

5 (a) DCDXCIX or CMXCIX

(b) 31 cards - four 1 cards and three cards each of 0, 2, 3, 4, 5, 6, 7, 8 and 9

Chapter 2 (page 17)

1

×	4	9	2	5	8	6	1	3	7
3	12	27	6	15	24	18	3	9	21
9	36	81	18	45	72	54	9	27	63
8	32	72	16	40	64	48	8	24	56
2	8	18	4	10	16	12	2	6	14
6	24	54	12	30	48	36	6	18	42
5	20	45	10	25	40	30	5	15	35
4	16	36	8	20	32	24	4	12	28
7	28	63	14	35	56	42	7	21	49
1	4	9	2	5	8	6	1	3	7

With practice, a time of just under a minute is possible. Answers do not need to be neat, just legible!

2 (a) 19

(b) 4

(c) 5

(d) 8 or 32

(e) 16

(f) 8 or 27

3 (a) 9

(b) $^-4$ and 4 or $^-5$ and 5

(c) 8 and $^-5$

(d) (i) $^-4 + 5 = 1$

(ii) $6 - 4 = 2$

(iii) $^-3 \times {}^-5 = 15$

(iv) $8 \div {}^-2 = {}^-4$

4 (a) $2 \times 3^2 \times 5 \times 7$

(b) 17 640

5 (a) 6 6.066 6.6 6.606 6.66

(b) 0.66

6 (a) 300

(b) 4

7

Fraction	Decimal	Percentage
$\frac{7}{20}$	0.35	35%
$\frac{1}{4}$	0.25	25%
$1\frac{1}{5}$	1.2	120%
$\frac{3}{20}$	0.15	15%

8 (a) $\frac{13}{50}$

(b) 0.65

(c) £3.20

(d) $\frac{27}{50}$

(e) £6

9 (a) 44

(b) 84 kg

(c) $\frac{1}{5}$

(d) 10 days

10 (a) £9

(b) 20%

(c) £32.40

(d) 28%

11 (a) (i) $\frac{5}{12}$

(ii) $\frac{7}{12}$

(b) (i) $\frac{1}{4}$

(ii) $\frac{4}{9}$

(c) (i) $4\frac{1}{12}$

(ii) $1\frac{1}{3}$

12 7.2 m

13 (a) 60° 120° 60° 120°

(b) parallelogram

14 (a) Brazilian 120 g, Honduran 80 g

(b) £20

(c) £3.00

Chapter 3 (page 33)

1 (a) Answer should be about 3000; answer should end in 5

(b) 2955

(c) Forgot about place value and didn't put a zero in the units column when multiplying by ten.

2 361

3 245

4 (a) 45 × 3 is 135

(b) 43 × 5 is 215

(35 × 4 is 140, 34 × 5 is 170, 53 × 4 is 212, 54 × 3 is 162)

5 The total cost of the items was £20.03 (3 pence more than £20).

Chapter 4 (page 42)

1 (a) 9.54

(b) 2.54

(c) 21.14

(d) 1.51

2 (a) 49.05

(b) 13.89

(c) 2.3765

(d) 19.4

3 (a) 34

(b) 60.3

(c) 8

4 (a) (i) 4.92

(ii) $^{-}1$

(b) 5.92

5 (a) (i) $\frac{40}{9 \times 9}$

(ii) 0.5 $\left(\frac{1}{2}\right)$

(b) (i) 0.490 183 973 3

(ii) 0.49

(iii) 0.490

6 (a) (i) 7.391 975 309

(ii) 7.39

(b) (i) 3.057 607 09

(ii) 3.058

7 (a) 26 points

(b) £115

(c) £135

(d) 360 ml

(e) 9801

8 (a) 25%

(b) 50 g

(c) £120

(d) 4.5 ha

(e) 08:20

Chapter 5 (page 57)

1 Answers vary.

2 43

3

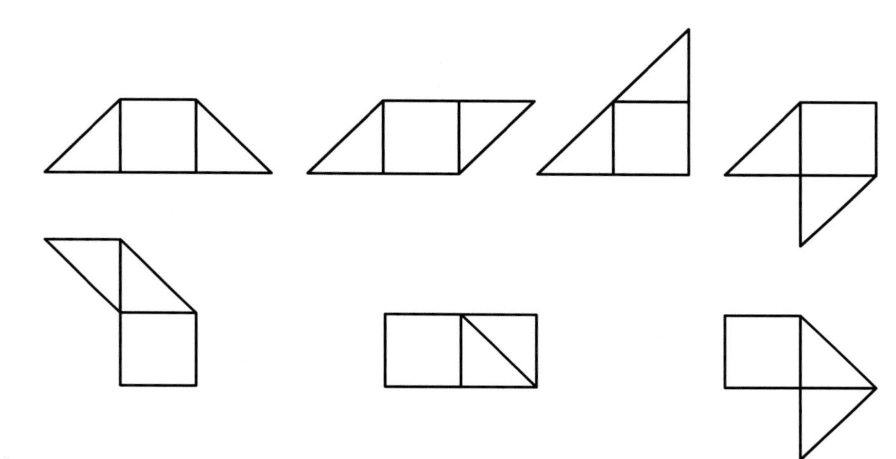

4 (a) (i) Brenda

(ii) Angus

(b) 2 minutes 19 seconds

(c) second

5 (a) 60 green balls

(b) (i) 10p

(ii) £10

(c) green

(d) equally likely

Chapter 6 (page 65)

1 27 (You probably found first that there are 6 diagonals from each vertex and then halved 9 × 6)

2 5 (You probably made an organised list.)

3 5.5 ($\sqrt{25}$ is 5 and $\sqrt{36}$ is 6, so it can be assumed that the square root of 30 is a little less than 5.5; 5.4 × 5.4 gives 29.16 and 5.5 × 5.5 gives 30.25 so, to 1 decimal place, the answer is 5.5)

4 76, 156

5 16 (You probably tried a simpler – shorter – maze.)

6 498

7 (a) $\frac{5}{12}$

(b) $\frac{1}{6}$

(c) 24

8 (a) £224

(b) £292

(c) 45 hours (13 extra hours)

9 (a) 2 cm

(b) (i) x cm

(ii) $x - 4$ cm

(iii) $x^2 + 4x - 16$ cm^2

(c) (i) 464 cm^2

(ii) 640 cm^3

10 (a) (i) 24.61

(ii) 26.20

(iii) 26.11

(b) Beatrice and Colin

(c) 85.56 kg

Chapter 7 (page 74)

1 (a) 7

(b) 28

(c) 8

(d) 4

2

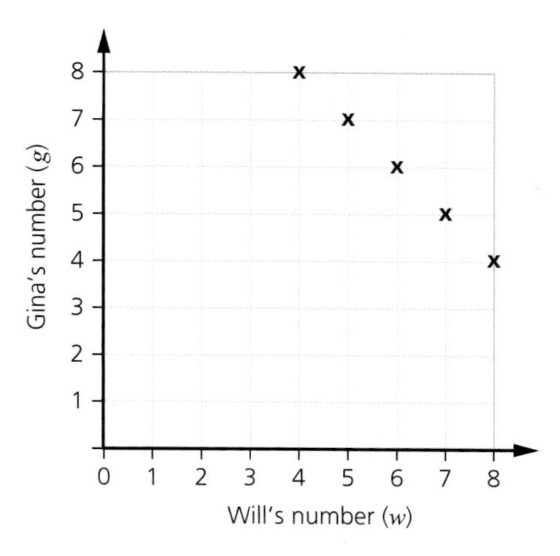

3 **(a)** 4

(b) 6

4 **(a)**

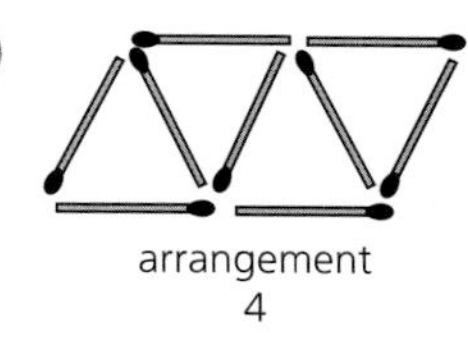
arrangement 4

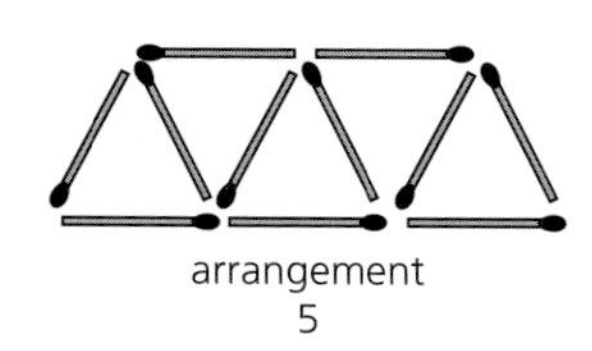
arrangement 5

(b) **(i)** 13

(ii) 21

(c) **(i)** yes

(ii) 201

5

(0)	1	2	(3)	4	**5**	(6)	7	8
(9)	**10**	11	(12)	13	14	**(15)**	16	17
(18)	19	**20**	(21)	22	23	(24)	**25**	26
(27)	28	29	**(30)**	31	32	(33)	34	**35**
(36)	37	38	(39)	**40**	41	(42)	43	44
(45)	46	47	(48)	49	**50**	(51)	52	53

6 **(a)** **(i)** points *A*, *B*, *C*, *D* plotted and labelled

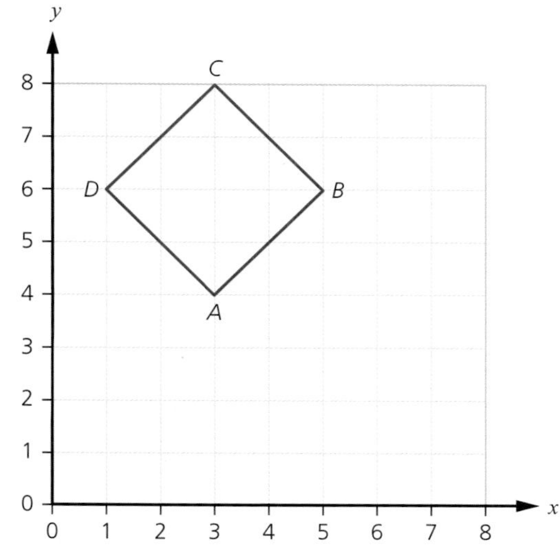

(ii) *ABCD* drawn; square

(b) isosceles triangle *PQR* with co-ordinates

Chapter 8 (page 83)

1 (a) $6k$

(b) $6k$

(c) $3k^2$

(d) $9k^2$

(e) $9c^3$

(f) $20t^6$

2 (a) $4a - 8$

(b) $6a + 15$

(c) $11a - 10$

(d) $10t + 15u - 2v$

(e) $e - 6f$

3 (a) $2rs + 5r^2$

(b) 0

(c) $5b + 23$

(d) $\frac{4}{5c}$

(e) $5b$

(f) $8n$

4 (a) $3(a + 2)$

(b) $2(2b - 5)$

(c) $2(a + 2b - 4)$

(d) $a(a + 2)$

(e) $3r^2 (2r - 1)$

5 (a) 12

(b) 8

(c) 18

(d) $^{-}30$

6 (a) $c + 4$

(b) $c - 2 = 2(c - 6)$ leading to $c = 10$

(c) 17

7 (a) (i) $y - 4$

(ii) $3y$

(b) $\frac{5y - 4}{3}$

(c) $y = 14$

(d) 10, 14, 42

8 (a) $d = 19$
(b) $w = 3$
(c) $c = 3$
(d) $a = {}^{-}1$
(e) $e = 2$
(f) $b = 6$

9 (a) $d = 12$
(b) $r = 20$
(c) $y = 8$
(d) $x = {}^{-}1\frac{1}{2}$
(e) $x = 24$
(f) $x = 0$

10 (a) x is 8.45 (or x is 3.55) gives y as 29.9975
(b) x is 3.55 (or x is 8.45)
(c) x could be 2 or 10 since 2(12 – 2) = 20 and 10(12 – 10) = 20

11 (a) 9, 6
(b) 10, 2.5
(c) 18, 29

12 (a) 18, ... , 24
(b) 36, ... , 64
(c) 26, ... , 50

13 (a) 10, 28, 82, 244
(b) 12, 30, 84, 246

14 (a) multiply by 2 and then subtract 1
(b) $T_n = 4n - 1$

15 (a) (i) $T_1 = {}^{-}1$, $T_{100} = 9998$
(ii) $n = 21$
(b) (i) $T_1 = \frac{1}{4}$, $T_{100} = \frac{199}{301}$
(ii) T_n approaches $\frac{2}{3}$

16 (a)

x	$^{-}3$	$^{-}2$	$^{-}1$	0	1	2	3
x^2	9	4	1	0	1	4	9
$2x^2$	18	8	2	0	2	8	18
y	13	4	$^{-}1$	$^{-}2$	1	8	19

(b)

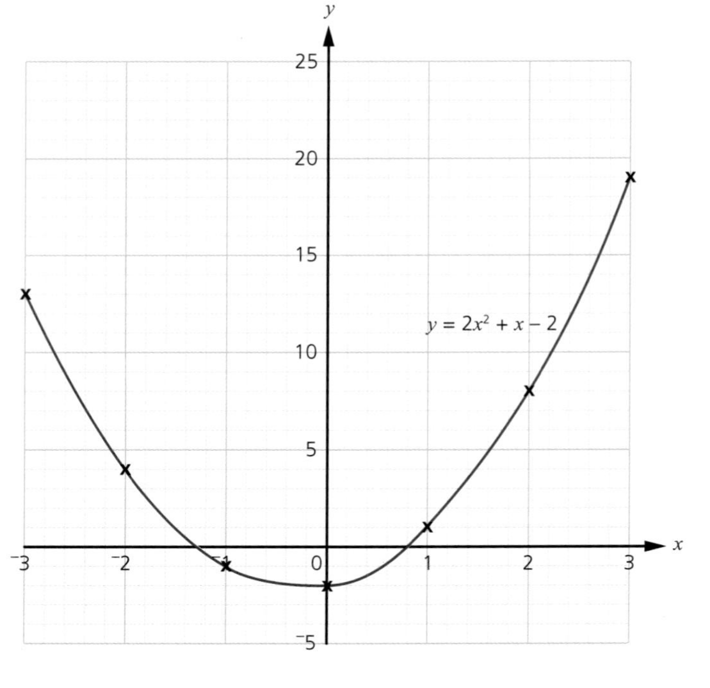

17 (a) (i) $m + n$

(ii) $m - n$

(b) $m + n = 20$

(c) Natascha is 7 years old ($m - n = 6$, so $2m = 26$, leading to $m = 13$)

Chapter 9 (page 103)

1 A 140 g

B 125 g

2 (a)

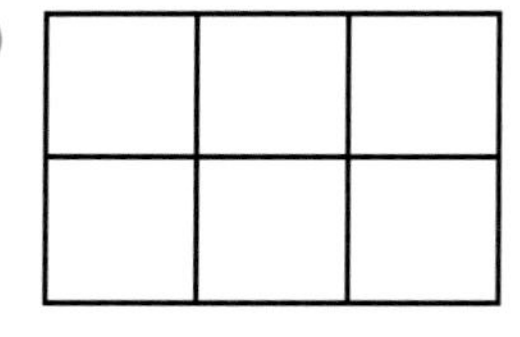

(b)

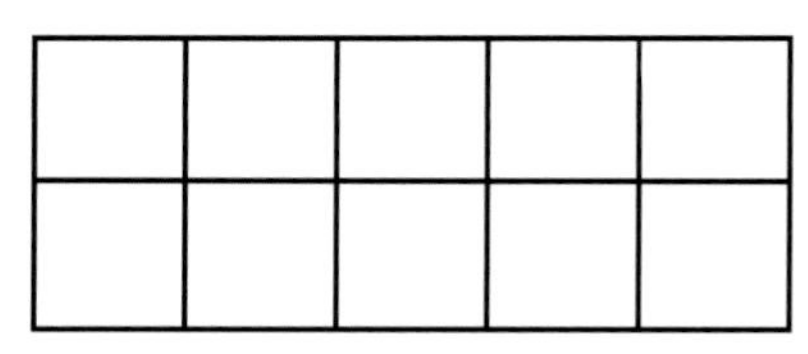

3 (a) 9 cm^3

(b) For example:

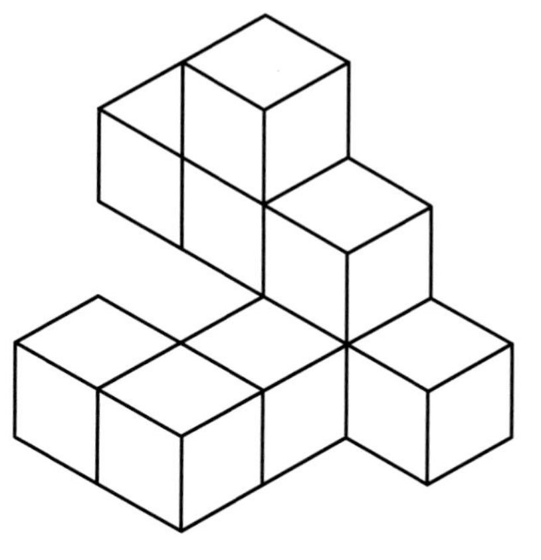

4 (a)

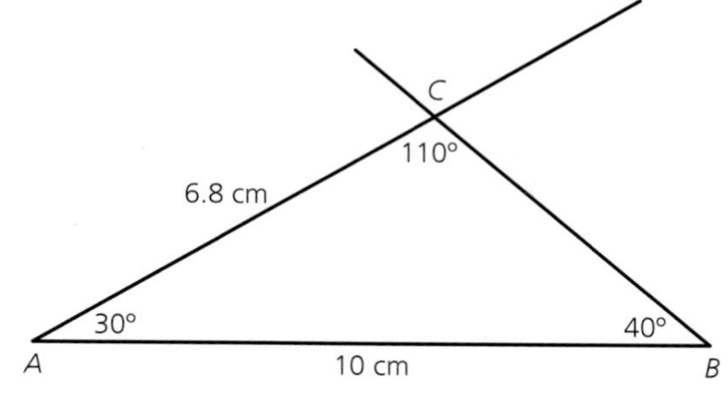

(b) (i) 110°

(ii) 6.8 cm

5 (a) congruent – the same shape and size

(b) (i) B

(ii) C

(iii) E

6 for example:

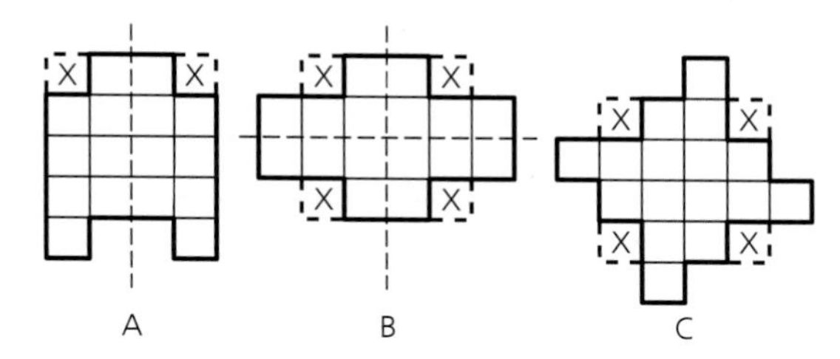

Chapter 10 (page 121)

1 Note: All answers are rounded. The official rules for sports usually give a range of measurements.

(a) (i) 70 mm

(ii) 40 mm

(iii) 25 cm

(b) (i) 160 g

(ii) 3 g

(iii) 120 g

(c) (i) 300 ml

(ii) 180 litres

2 (a) approximately 625 mm² (661 mm² – 36 mm²) or 6.25 cm²

(b) approximately 1250 mm³ or 1.25 cm³

3 (a) 8 minutes 24 seconds

(b) 2 minutes 6 seconds

(c) 7 seconds

4 A (a regular hexagon) 6 lines, rotation symmetry order 6

B (a rhombus) 2 lines, rotation symmetry order 2

5 $a = 72°$, $b = 36°$, $c = 72°$, $d = 54°$, $e = 54°$

6 (a)-(e) (i)

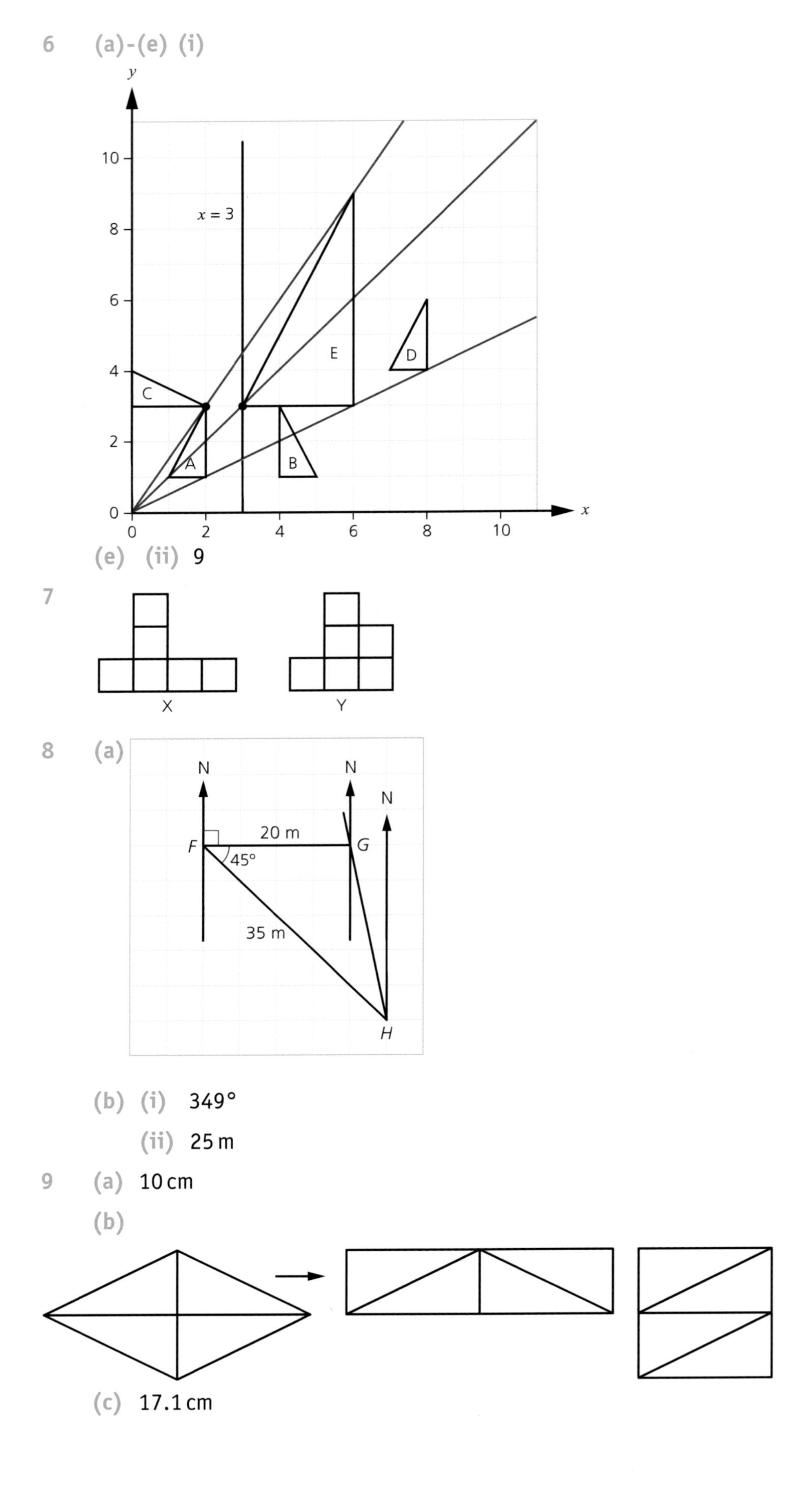

(e) (ii) 9

7

8 (a)

(b) (i) 349°

(ii) 25 m

9 (a) 10 cm

(b)

(c) 17.1 cm

Chapter 11 (page 144)

1 **(a)** **(i)** 13

(ii) 17

(iii) 13.5

(b)

Grade	Scores	Tally	Frequency
A	17–20	~~IIII~~	5
B	13–16	~~IIII~~ II	7
C	9–12	~~IIII~~ I	6
D	5–8	II	2
E	1–4		0

(c)

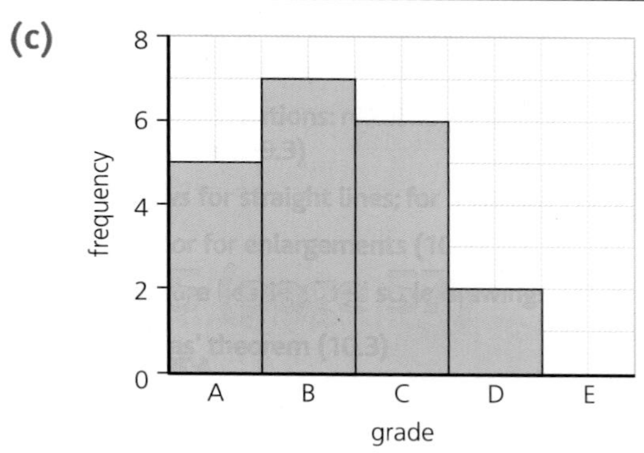

(d) 18 (90 ÷ 5)

2 **(a)** 1130

(b) 10 km

(c) 1330

(d) $1\frac{1}{2}$ hours

(e) 30 km

(f) 5 hours

3 **(a)** $2\frac{1}{2}$ miles

(b) about 1500 m

4 Answers vary. It could be a good idea to discuss your answers with a friend.

5 **(a)** **(i)** This was *not* a good strategy since there were insufficient trials.

(ii) no

(b) No – this die would *appear* to be biased towards 1

(c) A large number of trials is needed.

(d) The stallholder could:

- make sure that any biased die looks attractive
- make sure that other people cannot watch what happens.

Chapter 12 (page 156)

1 (a) $3\frac{1}{2}$

(b) 8

(c) $8\frac{1}{2}$

2 (a) 48

(b)

Marks	Tally	Frequency
40–49	III	3
50–59	III	3
60–69	~~IIII~~ III	8
70–79	~~IIII~~	5
80–89		0
90–99	I	1
	Total	20

(c) 60–69

(d)

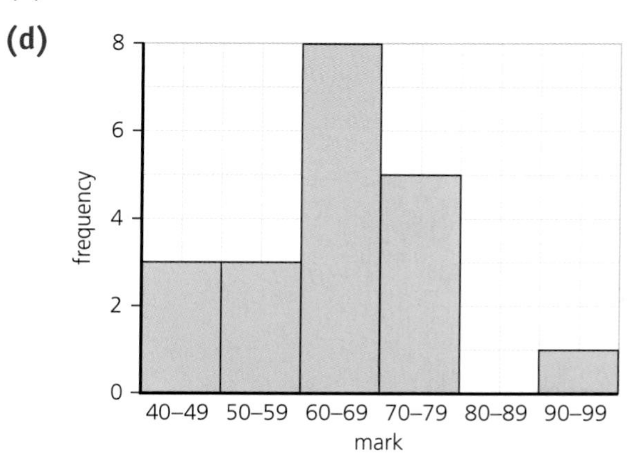

(e) The second paper was probably more difficult; only one pupil achieved more than 79% compared to 7 pupils in the earlier exam, etc.

3 (a) 2.5° each bulb; 36 daffodils (90°); 9 hyacinths (22.5°); 36 snowdrops (90°); 18 tulips (45°); 45 crocuses (112.5°)

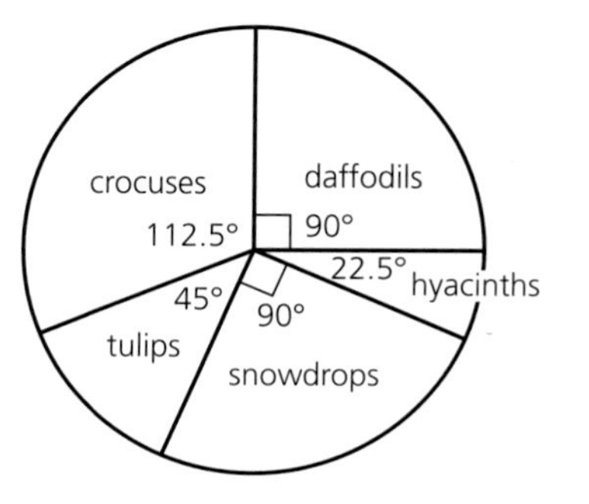

(b) 37.5%

4 (a) about 100 inches

(b) 175 cm

5 **(a)**

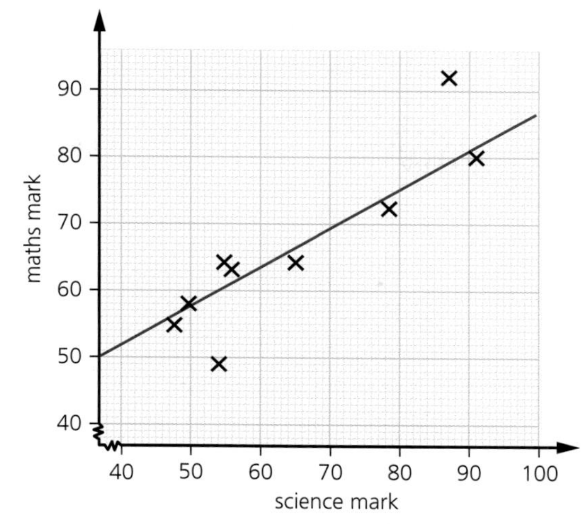

(b) Line of best fit – opinions may differ!

(c) Mathematics mark possibly about 64%

6 **(a)** **(i)** $\frac{1}{49}$

(ii) $\frac{24}{49}$

(iii) $\frac{8}{49}$

(b) $\frac{1}{48}$

7 **(a)** **(i)** $\frac{1}{22}$

(ii) $\frac{15}{22}$

(iii) $\frac{21}{22}$

(b) $\frac{5}{7}$

8 **(a)** **(i)** 41, 43, 45, 47, 49

(ii) 40, 45, 50

(iii) 42, 49

(b)

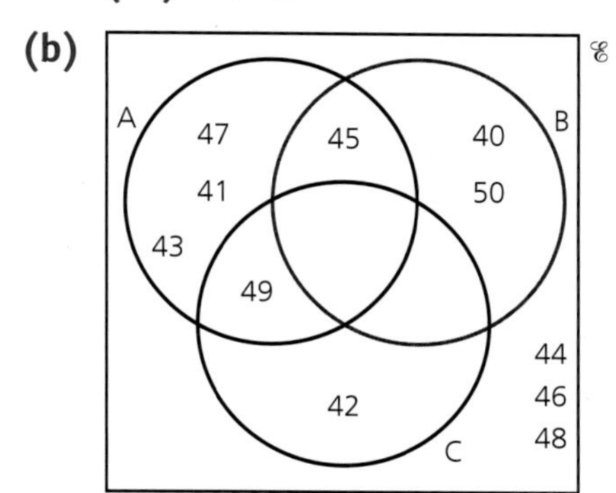

(c) 0